AF880013

Wildlife Rescue Rehabilitation and Restoration

The Journey of Rescue, Rehabilitation and Coexistence of our native wildlife

By

Olivia LoBalbo
State and Federally Permitted Wildlife Rehabilitator

Table of Contents

Introductions

Not so long ago, humans viewed themselves as caretakers of the planet, responsible for honoring and replenishing the gifts given to us by the Earth. People respected, nurtured, and revered the environment. However, over time, this perspective has shifted. Currently, a notable portion of the present generation perceives themselves as merely requiring resources from the Earth. They like to think that they're in charge, manipulating it solely for personal gain without balance. It's crucial to address and rectify this significant imbalance quickly.

There's a growing need for equilibrium. Talking about it isn't enough. To thrive as a species and ensure a sustainable environment for future generations, we need to advocate for the rights of the Earth and all its inhabitants. We must seek a harmonious coexistence between human expansion, our carbon footprint, and the biodiversity and beauty of our planet. I'm not suggesting a halt to progress. I recognize the impossibility of stopping human expansion and do not intend to attempt it. However, I advocate for moderation as a representative of the fauna and flora. Expansion doesn't have to be an all-or-nothing endeavor. For countless years, we've cohabited with our environment. Let's not sever that relationship now.

Let's talk about the importance of being good neighbors to wildlife. As human development progresses, it's crucial to recognize the inherent natural ecosystem that also requires our support in our fast-evolving world. Urban expansion fills the gaps in our environment, but it heavily relies on the existence of a thriving natural world. Being a responsible wildlife neighbor means actively striving for peaceful coexistence with the native fauna in your area while minimizing any adverse impacts on their habitat. It involves acknowledging the responsibility of owning land—vast acres or a simple balcony. Land ownership isn't just ownership; it's stewardship earned through hard work. Everyone, regardless of the size of their property, can contribute to nurturing a healthy environment.

Join me on this journey as we transform what was once "human space" into a rewilded, natural environment to restore vitality and harmony to our lives.

The starting point of this journey lies in comprehending unequivocally that the world owes us nothing. It's a challenging concept, one I grapple with myself. Life is a gift given to us but fleetingly capable of being withdrawn instantly. Our contribution to the environment should be our legacy. It's all any of us can do—contribute positively before departing. That is an eternal legacy if you've added beauty or kindness to this world. Why not make that the primary goal?

We undeniably owe a debt to the environment. This debt must be repaid to sustain the symbiotic relationship between humanity and nature. I aim to explore various avenues, ideas, and initiatives that can positively impact the environment. Each individual possesses the capacity to contribute significantly. Reflecting on our position in this world is an integral part of the human journey. It's crucial not to be disheartened by past actions or daunted by the enormity of the task ahead. Instead, consider your actions today to create a meaningful difference!

In our quest for the 'next big thing,' coming up with bigger and better than we had before becomes increasingly challenging each year. In the world that has given us so much magic and wonder, many people forget about the simplistic beauty of the natural environment. They overlook the serene beauty that is right here in nature—the origin of true magic. It's interesting how we aim to instill wonder and awe in children, often through what impresses them. Yet, children frequently find joy in the simplest natural phenomena—seed germination, fledgling birds taking flight, playful exploration by a creek, or scaling trees. Recalling moments from my childhood, the hours spent engrossed in video games lack the profound joy of reminiscing about time spent in nature. It's not just a memory. It's a feeling that has stayed with me my entire life. Nature provided a hands-on platform for learning, fostering a sense of being part of something grander than ourselves—a world teeming with magic. Despite our efforts, replicating this feeling remains beyond our reach. Regrettably, we've limited opportunities for our children to explore freely, discouraging tree climbing or playing in autumn leaves due to perceived dangers or bugs. As society increasingly fixates on the superficial world of electronics and screen time, mental health suffers, and overall happiness declines.

While acknowledging the value of electronics as tools contributing to our species' success, they aren't the only tools. Granting children the freedom to explore nature unhurried and at their own pace unveils an undeniable transformation. Witnessing their self-discovery and connection to the outdoors—that's true magic.

How to Use This Book

In the initial section of this book, we will explore strategies for preventing conflicts with wildlife and fostering a harmonious coexistence. The subsequent part will address contingency plans in case of unforeseen issues, providing a framework for stabilizing wildlife for 24 hours until it can be transferred to a suitable wildlife care facility.

This book focuses on native Virginia wildlife, but the principles and guidelines discussed can often be applied to other animals in different regions. Information about renesting, reuniting, primary care, and setups for sick, injured, or orphaned wildlife can be universally beneficial. Please remember that wildlife rehabilitators prioritize the well-being of native species and may not aid invasive species due to their potential threat to native ecosystems.

Moreover, it's crucial to acknowledge that hunting native wildlife is legal in most states. While there are regulations in place to protect these animals during baby seasons, hunting is a cultural tradition deeply ingrained in many societies. Respecting these laws helps maintain a balance between human activities and wildlife preservation.

To any aspiring wildlife caretaker, it's essential to comprehend the risks involved. Beyond physical injury, there's the possibility of contracting diseases from animals and the emotional toll of caring for an animal that may not survive or releasing a rehabilitated animal back into the wild. It's a demanding responsibility that requires emotional preparedness, considering that, ultimately, a wild animal's best interest lies in returning to its natural habitat. There's no such thing as a content, captive wild animal.

Chapter 1
Understanding Wildlife and Coexisting

Respecting Habitats and Natural Behaviors

Every species of animal has its unique ways of behaving in nature. It's essential to understand and honor these differences. To truly understand them, spend time in nature, observing, studying, and learning to appreciate their actions. Everything in nature has a purpose, and getting to know these animals helps us appreciate their role.

Take the white-tailed deer during rutting season, from October to late November, as an example. These deer display territorial behavior during this time, becoming notably more active. Unfortunately, this period also sees a significant rise in deer-related car accidents. By enhancing public awareness through education, installing more blinking warning signs for drivers about increased deer activity, and perhaps reducing speed limits in areas with high deer presence, we could potentially lower the occurrences of collisions between vehicles and deer.

Consider our native box turtles as another example. Many of these creatures come into rehabilitation centers injured due to encounters with cars and lawnmowers. Educating people about box turtles' natural behaviors could make a profound difference in saving more lives. During hot summer days, these turtles are most active early in the morning and after rain showers. They often forage in yards, seeking out newly emerged earthworms, a favorite food. Increased awareness of these active times could improve our efforts to check for turtles on roads and assist them in crossing safely. Simply checking yards for turtles before mowing can significantly impact and save lives.

Observing, studying, and understanding these animals deepens our connection and empowers us to safeguard their well-being. In wildlife rehabilitation, a crucial focus lies in preserving animals in their natural habitats and preventing clashes with humans from arising. We emphasize educating methods that exclude animals from human spaces and foster conflict resolution strategies.

Humane exclusion techniques for wildlife are designed to manage and deter wildlife from entering human-occupied areas without causing harm to the animals. These techniques aim to strike a balance between protecting human property and the well-being of wildlife. Wildlife conflict resolution aims to find a sustainable and ethical solution to the conflicts that inevitably arise between humans and wildlife. These can be as simple as competition for resources to the extremes like property damage. In some cases, humans can feel threatened for their safety due to wildlife. Effective wildlife conflict resolution aims to balance the needs and interests of humans and wildlife.

Educating local communities about the behaviors and needs of wildlife species causing conflicts is crucial. It's essential to raise awareness about the undesired behaviors while also highlighting the potential benefits of their presence. For instance, if a neighbor constantly feeds feral cats, deterring other wildlife, such as opossums, from nearby properties, is challenging. Educating the neighbor about the significance of individually caring for the cats and potentially providing information about trap-neuter-**rehome** programs would significantly reduce the population of outdoor cats. This step decreases the number of wild animals attracted to these feeding stations and helps manage the feral cat population. You could go on to explain the many benefits that Opossums bring to our yards and how remarkable they are.

Engaging the community is a pivotal step in comprehensive education. Encourage participation in community-led initiatives promoting wildlife coexistence. For instance, an HOA implementing rules about securing trash cans in garages at night, regardless of whether individuals have experienced wildlife issues, is a preventive measure. Such actions remove temptations for wildlife to become a nuisance in the neighborhood.

The significance of research and wildlife behavior monitoring often goes unnoticed. Here's the breakdown: Spend time observing animals in your backyard. The more you watch, the more you'll grasp their behaviors and interactions with your property and better comprehend the factors contributing to conflicts. Research to understand the wildlife in your area. Document your findings with photos and videos, sharing these insights to advocate for these creatures as living beings, fostering coexistence rather than framing them as mere nuisances.

Lastly, create a long-term management plan addressing ongoing conflicts to sustain human-wildlife coexistence. Develop a plan that considers their natural behaviors and provides pathways for successful cohabitation.

Chapter 2
Explore the Exclusion Techniques

This book offers a range of techniques to address wildlife-related issues. We will go over these techniques now to give you a better understanding of the tools:

Fencing and Barriers

Installing physical barriers like fences, netting, or mesh helps prevent animals from accessing specific areas. While effective for some species, it's not universally successful. Designing fences with adequate height or overhangs can hinder climbing and jumping. Additionally, incorporating barriers at the base of fences can deter animals from digging underneath.

Netting

Netting is often a popular choice to deter birds and small mammals from being destructive. However, it's crucial to select netting suitable for the specific species and use it carefully. Some types of netting can pose entanglement risks and lead to injuries for unintended wildlife. For instance, while installing garden netting to stop rodents from digging, people might unintentionally trap songbirds and reptiles. Achieving a balance between meeting human needs and respecting wildlife interests is essential in using netting effectively and responsibly.

If any wildlife becomes entangled in the netting, it's essential to contact your local wildlife rehabilitator for guidance on safely freeing the animals. Considering alternative netting options for the future might also be wise. Additionally, leaving soccer nets outdoors full-time poses a risk of entanglement for wildlife. To prolong the net's lifespan and ensure wildlife safety, consider bringing the nets indoors when not in use.

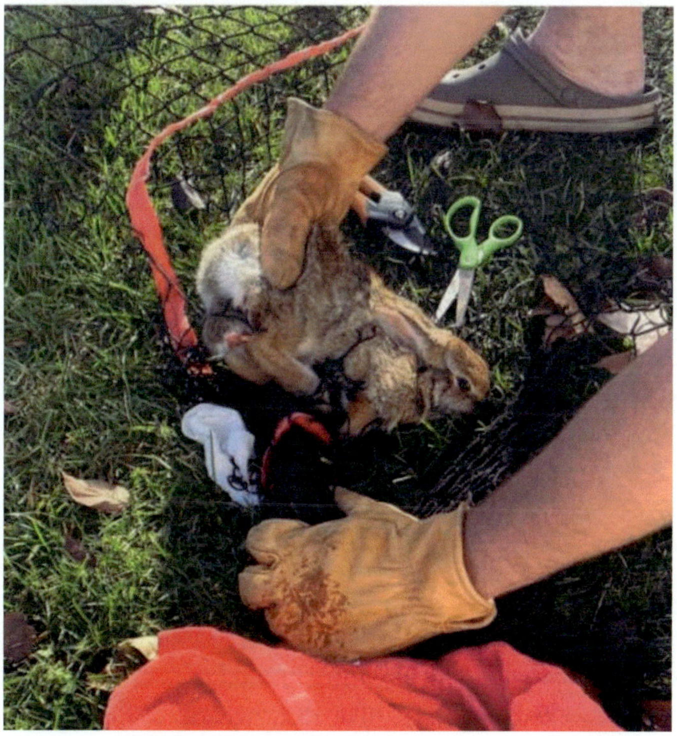

Sealing Entry Points and One-Way Exits

In situations where wildlife has entered your home, it's crucial to focus on humane methods to remove them rather than resorting to trapping and euthanizing. The initial step involves sealing entry points to prevent their return. This could include filling gaps, cracks, or holes, as well as addressing obvious access points like open windows or pet doors. Some companies utilize one-way exits, especially in cases like bats in the attic. By placing a camera during dusk to locate entry points, sealing these areas, and installing one-way exits, bats can leave at dusk but find re-entry blocked at dawn. This method has proven successful in keeping bats out of attics while allowing them to find new roosts, ensuring a positive outcome for the animals and homeowners. A well-placed bat box can also serve these animals, giving them a landing pad once removed.

Motion-activated lights and sounds

Harmless deterrents that prompt animal reactions. You can aid by observation with cameras. Using a Skatt Cat can, a motion-activated device releasing compressed air, is a simple yet effective deterrent. Originally designed to keep cats off countertops, it can be repurposed for gardens. For instance, in my vegetable garden, I've had success using it against a groundhog that loves munching on my Spinach. It's amusing to watch the groundhog attempt to take the food and get surprised by a puff of compressed air instead. Overall, it's a safe, effective, and occasionally humorous method to discourage unwanted wildlife.

Repellents

Repellents can be challenging. Most are ineffective—I've seen pictures of snakes near containers of "snake away" and curled up, sleeping near mothballs without any effect. Some repellents can also be unpleasant. Predator urine, although practical, requires widespread application across your yard to deter animals like squirrels and groundhogs. Plus, they need frequent reapplication to maintain their effectiveness.

Habitat Modification

Making environmental adjustments is highly effective in deterring wildlife. For instance, eliminating food sources like fallen fruit or unsecured trash cans can discourage animals from entering human spaces. Securing trash cans can be as simple as storing them in the garage at night or using locks to secure the lids, significantly if larger, more predatory animals like bears or raccoons are accessing your trash. You can easily find these locking mechanisms at local hardware stores or online platforms like Amazon.

Ultrasonic Devices

Ultrasonic devices pose a mixed success rate but are worth experimenting with if you aim to deter rodents or insects. These devices might be effective for smaller mammals such as mice and rats by emitting ultrasonic noises that are unpleasant to animals but generally inaudible to humans. However, they tend to be less successful in deterring larger animals like raccoons and foxes. They can also cause distress to household pets.

Tree and Plant Pruning

Trimming tree branches and shrubs away from buildings is crucial to prevent animals from using them as entry points. In one instance, I had released several orphaned squirrels on a property. A large oak tree in the front yard, a favorite spot for the squirrels, unfortunately, provided direct access to the roof of the house, and subsequently, they found their way into the attic. Despite trying deterrents like ultrasonic devices and sprays, we resorted to using a live trap in the attic, catching the squirrel, and sealing the entry point. I learned from this experience and now prioritize annual pruning of tree limbs, ensuring there are no direct access points for wildlife to enter my house, and I advise others to do the same.

Visual Deterrents

Employing moving or reflective objects like scarecrows, pinwheels, or reflective tape effectively deter unwanted wildlife. For instance, placing these items around your property can deter animals from entering. Moreover, affixing reflective tape and stickers to large windows is a preventive measure against birds colliding with them. Reflective tape comes in various patterns, making it a versatile addition to any home for deterring wildlife and safeguarding birds from window strikes.

When applying these methods, it's crucial to factor in the species involved and abide by local wildlife regulations and laws. Seeking advice from wildlife experts or organizations can guide optimal strategies tailored to your circumstances. Remember, when thoughtfully implemented, these techniques can be successful and often offer improved outcomes for humans and wildlife.

Chapter 3
Common Wildlife Scenarios to Remedy

Let's explore some common scenarios involving animals that we encounter and discuss practical strategies to prevent these issues from occurring. We will use examples from the many calls we receive to navigate helpful solutions and a better understanding of the animals.

Bats

Bats are a common concern in our area, often finding their way into attics, walls, or home siding. If bats have nestled into a snug spot that doesn't cause any disturbance to you, it's usually best to allow them to stay undisturbed. It is best to leave them be. Its great to have them outside and not enter your living spaces since they help control insects and benefit the ecosystem. If you notice their droppings around your home's exterior, a simple hosing can clean it off.

However, removing bats and preventing their re-entry is crucial if bats have entered your home. Installing a one-way exit valve allows them to leave but not return. Seal entry points before setting up the valve to ensure they don't return. Cleaning up their droppings must be done meticulously, often requiring professional assistance. Avoid using sprays or chemicals, as bats are susceptible to these substances. Providing them a chance to relocate humanely is essential, rather than resorting to extermination.

Timing matters, too—undertaking this process during warmer months gives bats the best chance to thrive. Disturbing them when they're dormant in winter can disrupt their ability to find food, water, or a new shelter before temperatures drop. Consider the season when dealing with these creatures. If they must go during the winter, reach out to your local rehabber to see if they're willing to overwinter the bats.

Picture: A Big Brown Bat in rehab who snuggled into Cookie Monster's mouth. Examples of them loving their crevices.

Bears

Spotting a black bear in your area can be pretty cool or scary, depending on how you look at it. With more houses popping up, bears are at home in our neighborhoods. Among the common concerns we receive are incidents of bears raiding bird feeders. A simple solution is to remove bird feeders during summer and early fall, providing seed only in the winter when necessary. Encouraging the growth of native plants is another way to reduce reliance on bird feeders. We're always talking about planting native stuff. Well, that could mean you don't need a feeder in the first place!

Keeping pets, especially dogs, indoors until the bear departs is crucial if a bear is in your yard. Sometimes, they're a bit too daring around bears. Employing tactics like loud noises—yelling, clapping, activating a car alarm—or using a foghorn can effectively scare them off and deter their return.

Securing chicken coops against predators is essential, and electric fencing may be necessary to protect against bear interference with beehives or gardens. While it's not pleasant to witness an animal being zapped, they learn quickly from these encounters and tend to avoid these situations in the future. It's better than thinking bears and humans can't live together, right?

Bobcats

Spotting a live bobcat is a stroke of luck, but it's wise to admire it from a distance. These cats usually prefer keeping to themselves and don't bother people or their pets. Nevertheless, keeping your cats indoors is the safest way to protect them from wildlife.

Most calls about these creatures involve car accidents or unnecessary panic because someone saw one. It's essential to know the real deal about these cats. They're larger than your house cat but smaller than a mountain lion. Instances of attacks on humans or livestock are extremely rare, but it's a good idea to secure your chickens in a predator-proof coop, just in case.

If a bobcat sets up a den on your property, it's usually for raising their young. Let them have their space, and they'll likely move on once their young are grown. Embracing these wild neighbors, giving them their distance, and appreciating their uniqueness is best. Take a moment to observe and appreciate how special they indeed are.

Coyotes

Coyotes are a common concern in our area, often affected by human expansion encroaching on their natural habitats, leading them into our neighborhoods. Being natural predators, you must ensure your pets are kept indoors and safe, especially avoiding leaving cats and small dogs unattended outside. They can also be territorial, so be careful with large free-roaming dogs. Otherwise, there's no real reason to fear these animals.

The key for Coyotes is the same as that of most other wildlife. Do not feed them, as this could diminish their natural wariness of humans. Ensure trash cans are secured, and avoid putting meats in compost piles (which isn't great for composting anyway). Also, it's worth listening to their distinct sounds—they can create a variety of unique calls, and when they gather, it can sound like a lively coyote congregation.

There's an old belief that they howl only after a successful hunt, but they're solitary hunters, primarily preying on small mammals like rabbits, rats, and squirrels. Their vocalizing is more about communication within their group and marking their territories to other coyote packs nearby. It's fascinating to take a moment, sit, and listen to their calls. You might even try recording them to notice the range of sounds they produce.

Deer

White-tailed deer pose a challenge in our area, mainly because of their size, and they're often labeled as 'nuisance animals.' I find this term inaccurate. While many believe there's an overpopulation issue, it's more about us encroaching on their habitats, seeing them as bothersome. It's ironic to get upset when we plant what they love and then get frustrated when they eat it. Consider using deer-resistant plants or, better yet, set up a trail camera for some wildlife entertainment if they visit your garden. Deliberately feeding them isn't a good idea; instead, planting native vegetation they enjoy can satisfy their needs without drawing them in.

We often receive calls about seemingly abandoned fawns. Adult deer typically leave their young hidden in bushes during the day, returning only at sunrise and sunset to nurse them. Despite appearances, they're usually not abandoned. Sometimes, if a mother is absent, other females in the herd may care for additional young.

The worst thing to do is to take a fawn found in the bushes home. It disrupts their natural care system and becomes heartbreaking for everyone involved.

Foxes

Foxes are incredibly stunning creatures known for their intelligence and social behaviors. Most calls we receive about them relate to cases of mange. Managing mange in foxes is tricky as the infection infestation is present in the environment, and healthy foxes usually fend it off. If you notice severe mange, it often indicates a secondary health issue. It's not as simple as treating them with a single dose of medication. While that might eliminate the mange mite, it doesn't restore the animal's health. They typically require more extensive treatments like antibiotics and additional care.

Besides mange cases, we often get calls from people who've spotted foxes in their yards. How lucky! Foxes are magnificent animals! They're much smaller than most people imagine, weighing about as much as an adult house cat. Despite common concerns, they're not much of a match for your cat or dog and won't run off with your children. These secretive creatures keep their distance and ignore you whether they're out in the woods or inside your fenced backyard. Their approach is simple: you're good if you don't bother them.

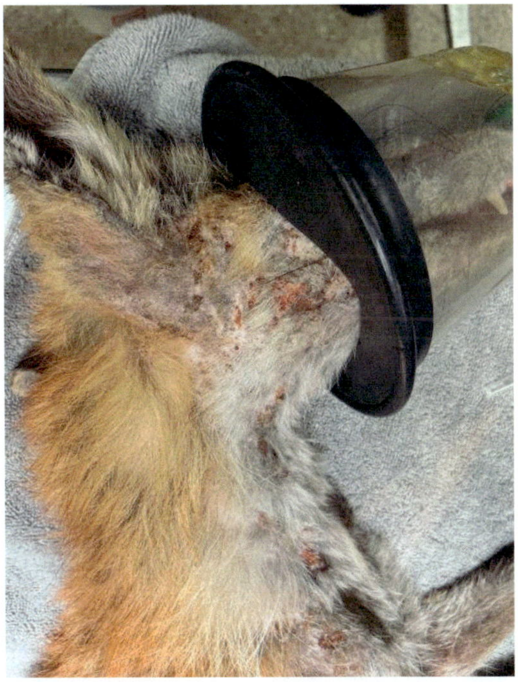

A picture of a sedated fox that is inflicted with mange mites. The skin is not just infected with mange but now has a secondary infection of the skin, eye infection, and severe malnourishment. This guy needs continued care, not just a quick dose of meds.

If you spot a fox out during the daytime, it doesn't necessarily mean it's sick or has rabies. Surprisingly, about as many domestic cats test positive for rabies annually in Virginia as foxes do. Sometimes, foxes are out during the day to keep warm—they don't hibernate like other animals. Breeding season typically occurs over winter, from December to February, with babies being born around March. So, no need to worry if you see them out and about doing their thing. Foxes are part of the natural landscape here.

Groundhogs

Groundhogs often face a tough time. Simply spotting them in daylight can worry people, yet these creatures are usually active during the day as part of their routine, not a sign of rabies or similar issues. They're easily frightened, seeing even small beings as potential predators. While they can contract rabies or carry parasites like brain-affecting roundworms, it's not common. Seeking advice from wildlife experts or rehabilitation centers is wise if you're concerned.

Many calls we receive involve groundhogs causing damage, especially under sheds or decks. When they have babies in spring and summer, it's crucial to let them be. Once the young ones leave, you can discourage their return by making the area less appealing for the mother, like using smelly items that resemble human presence. However, relocating them, especially during this time, isn't recommended as it can harm the babies left behind.

The best approach might be teaching these groundhogs that your garden isn't their territory. Some gardeners, like myself, plant extra crops specifically for them on the outskirts of your garden. I use cantaloupes as a distraction during summer. It's risky enough that once they grab one, they usually run off and only return when hungry enough to risk returning for more cantaloupes. If you're not keen on sharing your produce, a successful deterrent could be a fence around 3 feet high enclosing the whole garden. Scarecrows, fake owls, or compressed air devices can also help keep them at bay.

Mountain Lions

Mountain lions, or the lack thereof, often prompt numerous mistaken sightings—many reported cases turn out to be domestic house cats, with the occasional appearance of a bobcat. While mountain lions have an extensive habitat range, their population is remarkably low, and they are exceedingly elusive, seldom spotted in our area especially. Despite their scarcity, they endure a barrage of negative misconceptions. Sometimes, a mountain lion might wander closer to populated areas along waterways, catching the attention of humans. Our advice remains simple: maintain distance and give them their space. Similar counsel is extended to most callers dealing with various wildlife—ensure unintentional feeding isn't happening, typically preventing any issues. Avoid leaving pet food or meat in compost piles, as they might attract these animals.

Similar to other wildlife, there's a potential risk to livestock. Secure housing, particularly for smaller animals like chickens, is crucial. Larger livestock often benefit from protection by guardian dogs or even llamas! Deterring mountain lions through techniques like loud noises, flashing lights, or electric fencing helps set boundaries for wildlife.

Eliminating potential hiding spots is recommended for homeowners with extensive land concerned about mountain lions. Clearing brush piles or bushes near the house is the initial step to reducing cover for them. When all precautions fail, and a mountain lion remains a threat, the last resort is standing your ground! Mountain lions aren't monsters; they're essentially massive cats. Asserting dominance and making noise signals to them that you're assertive typically prompt them to keep their distance.

Despite their rarity, encountering these animals offers a chance to appreciate their beauty—a privilege reserved for only a few.

Opossums

Virginia Opossums easily tops my list as one of the most enjoyable animals to rehabilitate! They're adorable, easy to handle, and have healthy appetites. The only downside? They have a strong odor.

We receive numerous calls about Opossums, ranging from finding lone babies to adults raiding garbage cans or taking advantage of outdoor cat food. These crafty creatures often blend in with cats during feeding time. Let's tackle the simplest scenario first.

Baby Opossums should never be left alone without their mother. If you spot a small Opossum measuring less than 7 inches from nose to tail base, it's best to bring them in for rehabilitation. Sometimes, the mother might flee quickly due to injury or fear, leaving her young behind. Often, we receive calls about Opossums injured by cars, with concerned individuals discovering a pouch full of babies. Never lose hope! If you can get them to a rehabilitator, there's often a chance. Watch for these little ones, as they're around most of the year. There's nothing quite as endearing as a baby Opossum!

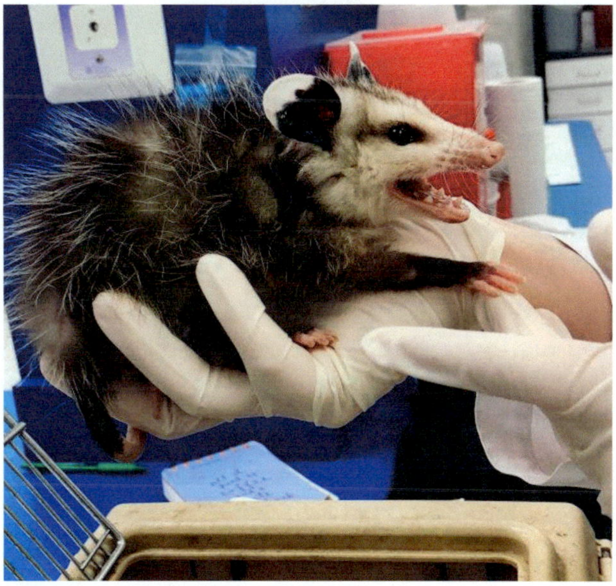

Opossums can find themselves in tricky predicaments. While it's wise to proceed cautiously, sometimes assisting them without involving animal control or rehabilitation services is possible. This depends on the situation, so reaching out to authorities or a wildlife rehabilitator is advisable before taking action. Simple issues like being trapped in a window well can be resolved by placing a branch or board for them to climb out. They're skilled climbers and will navigate their way out easily. If they're stuck in a garage, removing food and opening the door usually encourages their exit. Using a trail camera to observe their departure can be quite entertaining.

Despite the common perception that Opossums are unattractive, it's often just their "angry face." When they're comfortable, they're charming creatures. They're remarkably intelligent and display problem-solving abilities. You can discern their emotions from their little ears. They're likely at ease and curious if their ears are upright like Mickey Mouse. When their ears are flattened against their head with an open mouth and hissing, it's a bluff, but giving them space is essential regardless. These creatures are North America's only marsupials and offer numerous benefits despite their unusual appearance. What's better than having free pest control wandering your yard, dining on unwanted bugs, small snakes, and mice?

Raccoons

The first thing to note about raccoons is their intelligence! They quickly adapt, especially in suburban areas, thanks to their incredibly nimble hands that let them access things most other wildlife can't. This is the primary reason for most of our calls. The general advice for dealing with all wildlife applies to raccoons, too. If you leave out food, they'll likely show up. However, there are a few extra precautions for raccoons.

Securing garbage cans becomes crucial. Many find it helpful to tightly seal lids with bungee cords or invest in garbage can locks from platforms like Amazon. Dumpster diving is a common raccoon pursuit drawn by the scent of food. To get them out, placing a board inside allows them to climb out quickly. Afterward, close the top to avoid a return visit the next day.

Spotting raccoons during the daytime shouldn't immediately set off alarms. Though nocturnal by nature, it's not uncommon to see them during the day, especially when you leave out pet food, potentially attracting raccoons with their babies. Mixing pets with wildlife, including raccoons, is best avoided.

Patience is vital in their removal process if you have raccoons denning under your deck or in your garage. During spring or summer, they might be a mother with babies, and unfortunately, many baby raccoons end up without their mother after she's trapped and euthanized. Encouraging them to leave by making noise or being disruptive is advised. Installing one-way doors in attics can prevent their return once they've left.

Concerned about raccoon feces around your property? It's valid, as raccoons can carry dangerous intestinal parasites. Use gloves when cleaning up after them, and consider using repellents to keep them away, ensuring no food sources lure them back. They'll likely give you the space you need if they have no reason to visit. Remember, raccoons might be cute, but they're not suitable pets—spread the word!

Snakes

We contacted our local snake expert, Bill Crisp from K2C Reptiles, with questions about the top calls he receives about snakes. You can read more about Bill's work later in the book.

1) Snake identification - this is a daily text from customers looking for a quick ID on snakes they find in their yard or inside their home. Roughly 15-18 texts a day during peak snake season (May-August), but snakes in Virginia are primarily active between April and October and calls this year have continued into December. Once I receive the text of the snake in question, I will ID it and then have a follow-up conversation with the customer on any additional questions or services I can provide. Some result in just an ID, and some result in a removal.

2) Snakes in the Home - this occurs every week and consists of most times having to go out to the customer's house and remove the snake from basements, bedrooms, crawl space, etc... This starts with a quick text of the snake so I can ID it for them as I am heading out to their home. Snakes are sometimes stuck on glue traps or curled up in the laundry room behind the Tide container. I take this time to talk to them about not using glue traps and the damage that can be done, etc... Also, address possible entry points the snake may have used to enter the house.

3) Home Inspections - this results in a follow-up from #2 above. Once a customer has snake in the house typically results in a follow-up home/property inspection. This is a process of checking the inside and outside of the home, identifying entry points in the house and areas that need to be addressed to limit snakes from coming in or onto the property (i.e., overgrown landscaping, junkie yards that provide shelter for snakes, etc...). I will perform a full inspection inside and out and provide the customer with suggestions on how to remediate areas of concern. "

The takeaway from Bill's information is straightforward: Most people want to identify what they're looking at. Unfortunately, fear of snakes is common, often without a substantial basis. It's essential to recognize that most snakes prefer to avoid human interaction and will only strike when they feel threatened or cornered. Encountering a snake doesn't necessarily imply danger. Most snakes are harmless and shy away from confrontation.

Respecting snakes by leaving them alone benefits both humans and the environment. By allowing them to carry out their natural roles, we aid in maintaining a healthy and balanced ecosystem. Education about these creatures helps dispel unwarranted fears and encourages coexistence. While caution and awareness are necessary around snakes, it's equally crucial to acknowledge their significance and grant them the space they need to thrive. We can foster a harmonious relationship with these fascinating and ecologically essential creatures.

Skunks

Those little fellas can be pretty adorable if given the chance. They're not quick to spray unless they feel cornered or are startled abruptly. Contrary to popular belief, they're not as intimidating or troublesome as people tend to assume. The fear of being sprayed often prompts an overreaction, leading to the need to remove them from the property. Their vision isn't great, particularly during the day, as they're mainly nocturnal. Observing for a while is advisable if you spot them during daylight hours. Dawn and dusk sightings are normal. If you encounter a skunk needing help, approach slowly, speak softly, and move gently. Most calls we receive are simply due to someone wanting them gone. As with other species discussed in this context, that's not a fair outcome for any animal. Instead, removing resources and using deterrents is a better approach to keep them at bay.

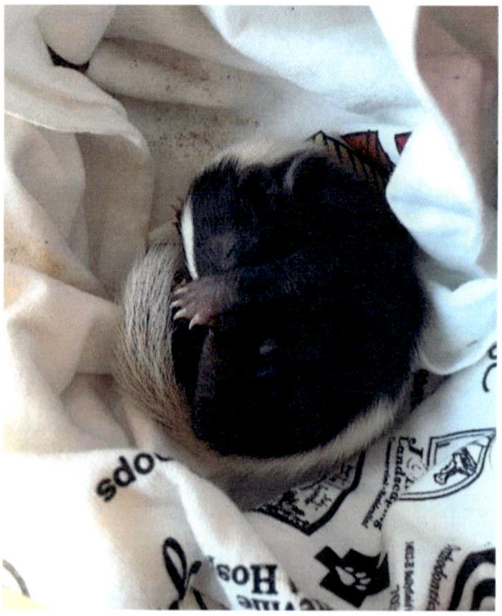

However, if they fall into man-made ditches like pools or window wells, they can be rescued with caution. Wearing gloves is essential as they can be carriers of rabies. It's wise to wear an old T-shirt if you inadvertently startle them. If you find skunk babies alone or near a deceased mother, using gloves and a towel to gently place them in a box is recommended for transportation to a rehabilitation center promptly.

In the frequent case of a dog getting sprayed, skip water rinses as they might exacerbate the smell. Opt for homemade rinses with baking soda, hydrogen peroxide, and dish soap instead of fancy de-skunking shampoos.

Remember, these creatures are generally peaceful and prefer not to spray any more than you'd like to be sprayed! You'll often find them with their noses to the ground, doing a great job eliminating garden pests you'd rather not have around!

Squirrels

Squirrels hold a special place in the hearts of wildlife rehabilitators, often being the first animals, they work with. Due to various issues adult squirrels face, there's a consistent need to rehabilitate their babies. Raising these young ones follows a straightforward plan, fostering a strong connection as they spend around three months in rehab from birth till their release.

Adult squirrels can cause problems in many suburban areas, especially where houses border wooded areas. One common issue includes squirrels raiding bird feeders, though there are solutions with squirrel-proof feeders. Some squirrels, being more agile and intelligent, might require inventive tactics to deter them.

When squirrels invade homes, it becomes a challenge. The best approach involves coaxing them out by making the environment uncomfortable. Squirrels have multiple nests, so removing one doesn't harm them. But preventing their return requires fixing the entry points in attic joints. Sometimes, squirrels get stuck in chimneys, necessitating methods to help them escape by guiding them out or providing a route using ropes. Capturing them is tricky; they're quick and may bite as their teeth, tinted orange due to iron reinforcement, are powerful. Handling them barehanded is never advised! Setting traps in the attic might be necessary in extreme cases. Once captured, it's crucial to ensure no babies are left behind before sealing entry points and releasing the captured squirrel nearby.

Orphaned squirrel babies are a common reason for calls to rehabbers. They can be identified easily by their black fingernails. Efforts are made to reunite them with their mothers, such as playing recordings of baby squirrel cries to attract maternal attention. Quick action is vital as these babies need warmth and feeding every few hours.

It's essential to recognize squirrels' significance in the wildlife community. Their role in seed dispersal and forest regeneration is crucial, making them integral to our ecosystems.

Songbirds

Songbirds are universally cherished and hold a special place in our hearts, prompting our dedicated efforts to protect them. One of the most frequent issues we encounter involves window strikes. Often, birds can't distinguish windows and collide, rendering themselves unconscious. It's crucial to seek a rehabber's help immediately. Even if a bird appears alert afterward, it's wise to gently place it in a dark, warm box and contact a local rehabilitator. Studies indicate that releasing them right after a window strike can lead to later fatalities. Employing window decals can be beneficial to mitigate window strikes. These decals, available in various designs from playful shapes to elegant patterns like pinstripes, increase bird visibility, reducing collisions.

Different scenarios demand different approaches. Nesting birds often pose challenges when they choose unconventional spots like front door wreaths. During nesting season, birds utilize various nesting spaces due to the depletion of their natural resources. While it might seem inconvenient when they nest in door wreaths, the solution is simple: wait for them to vacate the nest before removing it. In the meantime, use another door. Birds mature rapidly; the nesting materials can be safely removed once they leave the nest. However, until they depart, they are federally protected, and relocation is prohibited.

Specific concerns arise from birds nesting in porch rafters or fears about outdoor cats threatening them. Removing the cats resolves the safety issue, as birds have a right to these spaces. Ultimately, it's not up to us to determine where birds should nest—it's their choice. While it might be inconvenient for us, considering we've disrupted their natural habitats, our role now is to aid these creatures in their survival.

Chapter 4
Prevent Unintentional Harm

Try to limit the chemicals in the yard.

The use of pesticides, herbicides, and fertilizers in agricultural and residential settings within a shared watershed has had dire consequences for the survivability of our wildlife. For example, because turtles lay their eggs in the soil, they are especially vulnerable to exposure to these chemicals, leading to increased mortality rates and deformities among hatchlings. For instance, we had an unfortunate case with one of our resident Woodland Box Turtles, Yara, who was born without eyes—a tragic consequence of these environmental factors. There's been a troubling trend with more reptiles and amphibians showing up in rehab centers with genetic mutations, respiratory illnesses, and deficiencies in essential nutrients.

Yara the turtle, if you look closely, you'll notice she lacks eyes.

These issues should be seen as early warnings, akin to canaries in a coal mine, urging us to pay closer attention to the environmental impact on these animals. What affects them today could eventually affect us tomorrow. Research indicates that many chemicals we commonly use are linked to these mutations, illnesses, and deficiencies. Surprisingly, even essential pesticides used in our gardens can interfere with a box turtle's ability to absorb vitamin A, leading to severe respiratory infections.

Not long ago, we encountered a case where a pest control company had directly sprayed pesticides on a turtle. When the homeowner inquired, the company downplayed the issue, stating that the turtle would recover in a day or so. Sadly, the homeowner waited for three weeks before seeking help, and by then, the turtle's infection had progressed so severely that it caused irreversible damage to its eyes, ears, and parts of the nose. Regrettably, euthanasia was the only humane option. When we tried to educate the pest control company, they defended their actions by stating they had tested these chemicals on animals and seen no fatal outcomes. They seemed unaware of the severe swelling, discharge, and secondary infections these chemicals could cause. They did, however, note that even in humans, if it gets near their face, it can cause severe irritation.

It's now our responsibility as homeowners to take charge of this situation. We can choose safer pest control methods and advocate against widespread chemical spraying that poses harm. Each person should engage, even involving the HOA or county supervisors, as these chemicals are often applied without our explicit consent or knowledge.

We must remember that animals lack access to remedies like eyewash or fluids to flush out these pesticides. If these substances are causing such detrimental effects on our animals, it prompts the question: what are they doing to us? Indeed, The usage of pesticides in neighborhoods is a hot topic, with claims that these chemicals break down rapidly and have little effect on wildlife or human life. However, if something has the potency to eliminate all mosquitoes, it raises concerns about its impact on the broader environment.

Indiscriminately spraying plants can lead to what's known as pesticide drifts. This occurs when the wind carries the chemicals and ends in unintended areas. Encouraging neighbors to spray directly on individual plants can help control this. I once met a master gardener who used a milk gallon with a slit around the plant before spraying. This clever technique allowed the gallon to catch any residual pesticide drift. When using pesticides or if your HOA plans to spray for mosquitoes or similar pests, it's crucial to have prior notification. This information is essential, especially if you have children playing outdoors or plan to take your dog for a stroll. The potency of the spray is higher the more recently it's been applied, so being aware is critical to minimizing exposure to these chemicals.

This is a quick easy way to target spray while gardening.

Addressing the Issue of Outdoor Pet Cats: A Concern for Wildlife and Feline Well-being

As someone who adores cats, I've enjoyed sharing my life with many furry companions over the years, and each one has nestled its way into my heart. Cats often find themselves on the shorter end of the stick compared to dogs in various aspects. Unlike dogs with that unmistakable "lost without you" quality, cats are commonly believed to be naturally inclined to the outdoors and thrive on their own. However, my extensive experience with cats in vet clinics, shelters, and rescue work has taught me otherwise.

When cats venture outdoors, they're placed in challenging situations they're not equipped for. Their lifespan significantly decreases, and the risk of injury skyrockets. Sadly, when they fall ill or get hurt, they're less likely to be brought to a vet clinic. People tend to relinquish their responsibilities toward these outdoor cats, almost viewing them as akin to wildlife.

The dangers outdoor cats face are numerous: heightened exposure to diseases, vulnerability to parasites, risks of vehicle accidents, poisoning, attacks from predators and other cats, exposure to harsh elements, and even abduction. These risks escalate drastically when cat owners neglect vaccinations and fail to spay/neuter their cats. The risks multiply with breeding, posing threats not only to adult cats but also to the health of newborn kittens. The roaming behavior associated with breeding increases the chances of encounters with cars and roads. Potential health issues include viral infections like feline leukemia, feline aids, feline infectious peritonitis, upper respiratory infections, parasite exposure to fleas, ticks, ear mites, gastrointestinal worms, and ringworm. Local shelters can attest to the overwhelming number of cats needing homes, making it unfair to introduce more cats into the environment before addressing the existing population facing euthanasia in shelters daily.

Feral and indoor-outdoor cats are not native wildlife and should be regarded with the same care as dogs. They rank among the most invasive animal species globally. Introducing them to any environment is unwarranted. Evaluating the risks to both your cat and our environment is essential. Cats are efficient hunters, sometimes killing without necessity. Though an instinct, this behavior could be mitigated if cats were kept indoors and well cared for. Cats pose a disproportionate threat to native birds and small mammal populations. Alongside habitat loss, the menace of free-roaming cats, dwindling insect numbers, and increased use of harmful pesticides, our native songbirds are facing a steep decline. These heartbreaking casualties are a daily reality in rehabilitation centers, affecting not only our songbirds but also our bunnies, squirrels, chipmunks, and more. It's a harsh reality for these creatures, and it's our responsibility to better protect our native wildlife by keeping our cats indoors.

Lastly, the issue of feeding wildlife can create unintended dangers for both animals and humans. While it might seem harmless to leave food outside for pets like dogs or cats, it often attracts unwanted wildlife. This situation can pose risks as wildlife may carry diseases or parasites that could harm your pets or yourself. Encounters between pets and wildlife, like raccoons or foxes, can lead to conflicts and potential danger. For instance, an interaction where a dog attacks a wild animal may result in the wild animal being tested for rabies, which often means it needs to be euthanized. This puts your pet at risk of exposure to the virus, requiring quarantine and possible legal consequences if vaccinations aren't up-to-date. Rabies, once symptoms appear, is incurable and fatal. To ensure safety for everyone, including pets and wildlife, it's essential to feed your pets indoors and respect boundaries. This practice minimizes the risks of unwanted interactions and potential health threats for all involved.

Here are some helpful tips for keeping your cats indoors

Begin with the essentials: spay or neuter your cat to curtail their urge to roam. Feeding your cat indoors exclusively is crucial; it encourages them to associate home with nourishment, ensuring they feel like a family member. Make their indoor experience enjoyable by introducing a scratching post and catnip, which instills healthy habits. Toys like a simple cardboard box or a grocery bag can provide ample entertainment. Experiment with a laser pointer; it's a fun game for you and your cat. Grant them a lookout spot by a window so they can observe the outside world comfortably from indoors. If your cat tries to bolt for the door, gently redirect them without causing fear. Loud noises or aggression might make them reluctant to stay indoors. Make indoors appealing and welcoming to them.

Encourage positive behavior by rewarding them with their favorite treat when gently picking them up to prevent them from escaping. Reinforce the idea that indoors is a safe and pleasant space for them. Don't let factors like pregnancy or allergies dissuade you. With precautions like regular vacuuming, grooming, and handwashing, even allergy sufferers can coexist with cats. I speak from experience. Discuss safety measures with your prenatal doctor if you're pregnant and own cats.

As a last resort, if constant indoor living feels challenging, consider a breakaway collar with a bell for your cat. The bell alerts wildlife about the cat's presence, and confining your cat during dawn and dusk, when birds are active, helps mitigate encounters. A breakaway collar will ensure the cat doesn't get stuck on anything, causing a life-threatening situation.

If cats were presented with a warm home, regular meals, and lifelong care, they'd choose that over struggling outdoors for the occasional bird hunt. The argument that "hunting is in a cat's instinct" doesn't necessitate throwing them outside. Satisfying their hunting instincts can be achieved indoors by providing toys and enrichment, just like responsible pet ownership entails. Claiming it's natural for cats to be outdoors is a weak excuse. There is nothing natural about having cats outdoors in our environment!

Believe in the process! Transitioning an older cat indoors is possible and gets easier over time. Remember, as a cat owner, you bear responsibility for their care and actions. Every animal they harm, every kitten they produce, every disease they spread—this responsibility falls on you. Bring your pet indoors; it benefits you, the wildlife, and your cat.

Window strike preventions

Preventing window strikes presents a challenge, particularly when certain bird species seem prone to colliding with windows. Several reasons contribute to these incidents. Sometimes, it's as simple as birds failing to perceive the presence of glass, while other times, they may mistake their reflections for intruders and defend their territory. These collisions become even more common during migration periods. Regardless of the cause, numerous birds end up in rehab facilities annually due to window strikes, leading to preventable deaths. Implementing a few straightforward measures can make a significant difference.

Step one involves identifying the window posing the highest risk. If birds have struck a particular window before, it's likely the primary concern.

Step two entails strategically placing either purchased or homemade window clings to deter birds from colliding with the glass. These decals offer visual cues that help birds recognize and avoid the window. They come in various designs, accommodating different home styles.

Step three emphasizes conducting periodic perimeter checks during peak migration times to locate and attend to injured or deceased birds. Switching to a different pattern or design might be necessary if collisions persist despite using window decals.

Making wildlife reliant on humans

Assisting wildlife can feel like an instinctive responsibility, leading us to provide food, shelter, and care for their daily needs. However, while offering water is a good start, going further by providing daily meals and shelters can inadvertently do more harm than good. It is crucial to recognize the pitfalls of making animals reliant on humans for their necessities.

There are several issues with this approach. Firstly, what happens when you're unable to provide food one day? Missing a meal can disrupt an animal's routine and survival. This level of constant support is unsustainable, and even if feasible for one animal, it can attract more, leading to overpopulation beyond the environment's capacity. Moreover, this reliance on humans can foster a false sense of security and abundance, prompting animals to reproduce more than their habitat can support. Additionally, habituating wildlife to human interaction poses risks. While you may treat them kindly, not everyone does, jeopardizing their safety in the long run.

It's essential to reconsider taking wild animals under your care and fostering familiarity. While it might feel rewarding, we must prioritize their well-being. Our interactions might make us feel special, but assessing the impact on their natural behaviors and survival in the wild is crucial.

Chapter 5
Provide Natural Landscapes

Familiarize yourself with your natural environment.

Beginning your journey toward land restoration involves understanding your local area's native plant life. It is crucial to learn about your property's history and the natural habitat it once offered before development. This knowledge lays the groundwork for restoring the land to its original state while harmonizing it with your home and family. Familiarizing yourself with your land and promoting biodiversity are the initial steps in reintroducing wildlife to your surroundings. Habitats and ecosystems are intricately linked, forming intricate food chains and connections between flora and fauna. This interdependence maintains ecological stability, where each element relies on others to sustain it. Disturbing or removing a single habitat or aspects of a habitat can trigger a chain reaction, unsettling the entire ecosystem.

Let's delve into attracting wildlife, starting with insects—an essential step in transforming your yard. Enhancing biodiversity and offering diverse food options are critical in drawing more wildlife. Whether you aim for various songbird species, larger mammals like opossums and deer, or numerous butterfly species, abundant food is crucial. Our native pollinators not only serve as food sources but also provide habitats. Large native trees such as oaks play a pivotal role, supporting a multitude of insects, bird species, small mammals, and even larger wildlife. If your yard can accommodate an oak tree, consider it a natural wildlife hotel.

Various animal and insect species require specific sustenance, shelter, and reproduction habitats. Disrupting these habitats can lead to the decline of species dependent on them. Even seemingly small actions like replacing native plants with ornamental ones can have severe repercussions on the surrounding insects and animals, upsetting the delicate balance and causing widespread issues. A minor alteration can lead to significant problems in the ecosystem.

I strongly advocate for homeowners to explore Douglas Tallamy's "Bringing Nature Home." Tallamy emphasizes the significance of planting native pollinators that attract specific native insects and creating wildlife corridors to facilitate safe animal movement. These native insects, like caterpillars, constitute a crucial protein source in a bird's diet, particularly during their early growth stages. A shortage of such food can hinder birds' breeding success, a phenomenon observed nationwide. The decline in migratory bird populations is primarily attributed to habitat loss. This underscores the critical need to rewild residential spaces, fostering a natural habitat and providing native plants for insects. By doing so, we can bolster essential backyard bird populations. Furthermore, native plants and trees are well-suited to local soil and weather conditions, requiring minimal maintenance and thriving with reduced water usage and no need for fertilizer—a beneficial extra.

Douglas Tallamy's book has sparked a collective effort among homeowners, fostering a movement toward establishing the Homegrown National Park. I highly recommend checking it out and adding your home to this expanding initiative. You can start right away by researching native plants in your region. Numerous seed-share programs and native plant nurseries offer delightful additions to your outdoor space. Take the first step today by planting a few in your yard.

In addition to native plants, it's enjoyable to cultivate species that attract the animals we enjoy having around. When we deliberately plant for specific species, we can often encourage them to dine on the native plants in our yards rather than the ornamental ones we've chosen purely for their aesthetics. For instance, to deter cottontails from munching on your tulips, consider introducing some of their favorite native plants, like dandelions. Explore the information below and contemplate planting or allowing some native plants that naturally emerge in your yard. The wildlife will certainly appreciate it!

Composting your leftovers - Supporting wildlife from the soil up.

Wildlife thrives within a healthy ecosystem. Establishing a harmonious coexistence with your wildlife neighbors begins by ensuring your land is as nurturing as possible. It's a process that begins at the ground level, and even small actions can significantly impact the lives of numerous plants and animals. Here is some helpful information to help you get started.

Without composting, food scraps, yard clippings, newspapers, and cardboard often end up in landfills, squandering valuable resources. These materials, essential for a compost pile, play a crucial role in benefiting our ecosystem and gardens. Compost piles attract a variety of animals, from tiny creatures aiding in decomposition to larger ones like skunks and chipmunks that feed on insects found in a pile. Woodland box turtles might even hibernate or lay eggs in your compost.

Starting small is critical—use any container or designate a corner in your backyard for your compost pile. Choose a convenient, easily accessible location; however, keep it visible to stay mindful of its maintenance. Placing it directly on soil facilitates drainage and speeds up decomposition.

Gather cardboard, paper, toilet paper rolls, envelopes, and paper towels as "browns," Mix these with grass cuttings and food scraps, known as "greens." Both browns and greens are essential for the decomposition process. To prevent contamination, avoid adding cooked foods, meat, fish grease, dairy, diseased plants, or pet waste. Maintain the right moisture balance by watering if it's dry and adding more browns if it's overly moist. Then, allow time for the magic to happen.

While the compost matures, consider setting up a trail camera to observe the diverse animals frequenting the pile for food, shelter, and even as a nursery! Over time, this composting process will yield nutrient-rich soil, benefiting numerous wildlife neighbors in your area.

A little fresh water can go a long way

Promoting wildlife welfare by providing water sources like bird baths or shallow dishes is a cause I strongly advocate for. In hot, dry Virginia summers, a sip of fresh water can be a life-saving difference for animals. However, offering clean water can spark controversy. Some neighbors might not appreciate the increased animal activity on your property, and specific homeowner associations (HOAs) might even prohibit it. Checking local regulations is crucial. If more animals gather around your water source, informing neighbors about your initiative can prevent unnecessary concern. Educating them kindly about the situation helps alleviate fears, especially if more significant, nocturnal creatures like foxes or raccoons might visit in the evenings.

Furthermore, offering fresh water comes with responsibilities. Selecting an appropriate dish for various species is essential. A shallow pan with added rocks to prevent drowning suits multiple visitors. Changing the water daily is vital. Stagnant water breeds diseases and attracts mosquitoes. Regularly dumping and refilling with fresh water maintains animal health.

Weekly cleaning of the watering dish using a diluted bleach solution—nine parts water to one part bleach—ensures hygiene and prevents disease transmission. In case of disease outbreaks in your area, especially avian diseases or rapidly spreading ones, it's prudent to temporarily remove the water source to prevent potential infections among animals.

Animals shouldn't have to either if you wouldn't drink from the dish. Prioritizing their health and safety is vital to supporting wildlife.

Bird boxes, bat boxes, bug hotels, toad rest stops.

Supporting the individual animals and nurturing your yard's ecosystem is a fantastic objective! A simple starting point is understanding the current inhabitants, identifying what you want to support or discourage, and considering any new animals you'd like to attract. Visualize your desired outcome: perhaps it involves the tranquil melodies of birds, the enchanting sight of butterflies fluttering in a field, or the captivating spectacle of bats darting around at dusk. Take a moment to articulate these aspirations by writing them down; transforming your vision into tangible goals is the initial step toward making this dream a reality.

Constructing nesting boxes and providing alternative habitats like birdhouses and bat boxes invites animals to settle and nest in areas separate from human structures.

The significance of trees should never be underestimated. Apart from native pollen providers, hardwood tree species offer fantastic opportunities for wildlife to find shelter and homes.

Regarding specific habitats like insect hotels, bat boxes, birdhouses, and squirrel boxes, their necessity depends on your yard's overall situation. Providing substitute homes becomes crucial for displaced animals in neighborhoods where trees have been depleted due to construction. These homes might be temporary until the regrowth of native plants and trees fosters natural habitats.

Starting small with an insect hotel is a charming and functional step when adding to your yard. These structures attract insects, providing shelter, but ensure you distribute them adequately to avoid making it easy for predators to prey on concentrated insect populations.

Bird boxes are another delightful addition, offering spaces for birds to nest and raise their young. However, it's essential to consider potential predators like hawks and adjust the placement or find tree cover for the boxes accordingly. Simple additions like overturned terra cotta pots in shaded areas can attract toads, small amphibians, and baby bunnies seeking shelter.

Structures like insect motels or bat boxes might not be necessary in yards, maintaining a natural balance. Instead, identifying species to support and preserve existing habitats or natural elements becomes more critical.

Supporting the wildlife already present in your yard through actions like leaving brush piles or retaining dead tree limbs for bat roosts can make a substantial difference to these animals' survival. While these acts may seem insignificant, they can be life-saving for the wildlife you support. Every effort, no matter how small, contributes significantly to their well-being.

Chapter 6
What to Remove from Your Environment

We've covered many additions that can benefit wildlife in your yard, but let's focus on elements we can remove to significantly improve animals' lives.

Chemical pesticides and herbicides

As we've discussed previously, pose significant issues. Their scent alone can repel wildlife, harming their health and disrupting the ecosystem. Simply discarding these chemicals isn't enough; proper disposal is crucial, and the labels often provide instructions for safe disposal methods. Moving forward, seeking chemical alternatives for pesticides and herbicides is vital. A little manual effort can significantly reduce the need for these chemicals—I am committed to banning pesticides and herbicides from my garden. I spend a few minutes each morning pulling budding weeds and removing grubs threatening my vegetables. Despite the small time investment, the payoff in the end is invaluable!

Removing Introduced Plants

You might have assumed that planting native species suffices, but unfortunately, that's not the case at this stage. To safeguard the native plants you've introduced, removing non-native plants from your yard is crucial. Invasive plants fail to offer our indigenous wildlife the necessary food or shelter. Therefore, finding the time to identify and eliminate these non-natives is vital! Though it may demand time and effort, consider it a bit of extra exercise that pays off significantly in the end!

Once more, let's reference Douglas Tallamy and his advocacy for fewer lawns and more native plants. Overdoing a meticulously manicured lawn, aiming for uniformity, strips it of any natural value. Allowing the lawn to dominate most of your property further diminishes its usefulness. Incorporating elements such as bushes, shrubs, and native plants into your landscape offers benefits while gradually erasing this distinction. An added perk? With each addition to the landscape, you'll find yourself mowing a bit less!

Trash

Frequently, people voice concerns about rats, leading to complaints about snakes, eventually followed by raccoon grievances. This sequence is a clear indication that there's accessible trash somewhere! Rodents don't materialize out of thin air; their presence indicates they've discovered a food source and a cozy dwelling. By eliminating these comforts—often in the form of trash piles or unsecured garbage cans—you can rid your space of rats, deter snakes, and discourage raccoons. It's a cascading effect. Tending to your yard involves more than just eliminating an eyesore; it's about removing potential attractions, keeping them clean and uninviting unwanted visitors.

Fallen fruits and vegetables

There's often a misconception that our gardens can inadvertently serve as an advertisement to wildlife, attracting unwanted pests such as insects, rodents, or larger mammals like groundhogs and raccoons. Ensuring to promptly pick up any overripe fruits or fallen vegetables can significantly reduce the presence of pests in your garden. You cultivated those plants for a purpose—harvesting them for your consumption. If you don't, there's a good chance an animal will take advantage of the opportunity.

Bird Feeders Overnight

Leaving bird feeders out overnight doesn't benefit the birds; instead, it often invites nocturnal animals like raccoons to feast on the seeds, causing a mess and potential damage to the feeder. Raccoons are remarkably cunning creatures! Once scattered, the seed attracts other wildlife like foxes, bears, and deer, making your sanctuary chaotic. The remedy is simple: during months when bird feeders are in use, bring them indoors in the evenings. Consider your backyard an extension of your home's interior—just as it needs attention and closure at day's end, like doing dishes and switching off kitchen lights, the yard requires the same care and tidying up.

Remove falling hazards for wildlife.

A fellow rehabber once faced issues with wildlife getting trapped in her pool. Incorporating straightforward wildlife steps into your pool setup is an inexpensive and effortless solution to prevent such incidents. These steps aid even the smallest creatures, like toads and snakes, all the way up to larger ones, like skunks, that might wander in but leave you hesitant to intervene due to fear of being sprayed. Do yourself a favor and add wildlife steps to your pool before it's an issue! Additionally, installing covers on window wells can prevent future incidents, saving you and the animals from unnecessary stress and potential harm. These covers are available for purchase online.

Outdoor Lights

Let's consider minimizing outdoor lighting, especially when not in use. These lights can disrupt the natural behaviors and rhythms of our nocturnal animals. While certain bulbs are less likely to interfere with wildlife, opting for motion sensor lights for your exterior is a better solution. This way, your home's exterior remains dark as nature intended at night, only lighting up when someone passes. If safety concerns arise, it's important to note that adding constant exterior lights may aid burglars in seeing better without necessarily deterring them. However, a motion sensor light that surprises someone when activated might give them a start.

Too much noise

Finally, excessive noise or disturbances outdoors can significantly drive wildlife away from their natural habitats. While our intention might sometimes be to discourage animals from lingering in our yards—something easily achieved by playing music continuously, for instance—this can have the desired effect of keeping wildlife at bay. However, if you aim to attract wildlife back into your yard, continuous noise blaring on your outdoor speakers can disrupt the simplest interactions among species like songbirds. Consider minimizing outdoor noise. Take a moment to listen to the natural sounds of nature and appreciate them for what they are.

We touched previously on wildlife conflict prevention, but unexpected situations can arise. This section will cover distinguishing normal behavior from abnormal in various species, understanding the developmental stages of young animals, and guiding those encountering sick, injured, or orphaned wildlife. The primary objective is to ensure their survival for 24 hours, not raise them long-term. For those considering becoming a licensed wildlife rehabilitator, reaching out to your local wildlife resource organization is the first step, or you can order A.E.R.O's online curriculum based on individual species by going to our website,

www.aeroanimalrescue.org

Chapter 7
Easy Reference Pages

The initial step in offering assistance is to assess the situation. Many people may misunderstand their behavior when encountering a wild animal for the first time. If you're unfamiliar with these creatures, it's easy to jump to conclusions about their actions, sometimes assuming they need help when they don't or overlooking signs of distress.

Upon witnessing an animal in apparent danger, the immediate urge is often to rush to its aid. While this instinct is understandable, pausing and evaluating the circumstances is crucial. This evaluation isn't just for the animal's benefit but also for your safety and adherence to regulations.

There are specific safety protocols to be followed and governmental guidelines dictating what can and can't be done to assist wildlife. For instance, certain species like coyotes and adult white-tailed deer in Virginia fall under nuisance wildlife, restricting intervention by state law. This limitation may be disheartening, but recognizing the reasons behind such regulations is essential. What's good for some isn't always good for all. No state agency will look at an animal as an individual but instead as a whole group that coexists with others.

1. Wild animals don't respond well to restraint. Firmly holding or trying to cuddle them might result in rejection from the animal, particularly with orphaned babies. Our human instinct to cuddle conflicts with their natural behavior. If a baby animal is resisting, nipping, or biting, it's a sign that our approach needs adjustment. Comforting wildlife is essential, but imposing our version of comfort may not benefit either us or the animal. Instead, observe the species and mimic a mother's behavior toward her young. For instance, a baby squirrel won't typically be rocked to sleep by its mother, so the young one may not welcome such actions.

2. Every sense matters to a wild animal—touch, sight, smell, taste, and sound. For an adult animal, especially when sick or injured, overwhelming these senses can cause distress. Avoid using too much perfume or playing loud music, as it can stress an animal unfamiliar with such stimuli. Consider these sensory aspects while interacting with wildlife.

3. Never underestimate a wild animal. Assume that any animal you handle might carry intestinal parasites or zoonotic diseases. Be cautious, as animals with teeth or beaks can bite unexpectedly. Plan for contingencies if the animal escapes, as catching it again might be challenging. Being mindful of these possibilities will save you trouble in the future.

4. All animals, regardless of their health status, desire freedom. When given the chance, they'll readily choose freedom over confinement. While it's not always practical to allow wildlife unrestricted movement, consider that forcing an animal into captivity against its will isn't in its best interest. Sometimes, humane euthanasia might be a more compassionate choice than confining an animal that yearns for freedom.

EASY REFERENCE PAGE

1. A general guideline suggests that most baby mammals can comfortably retain 5-7% of their body weight in their stomach. It's advisable to prioritize keeping an animal hydrated over attempting to feed it, as incorrect food could be detrimental. On top of that, feeding styles will vary from species to species. Some may need to be tube-fed vs bottle feeding. Never assume information found online about wildlife nutrition is accurate or suitable. It's crucial to consult a local rehabber with knowledge of the local wildlife feeding requirements, who might advise against feeding the animal anything. One size does not fit all in formula feeding.

Additionally, be mindful that oral fluids could accidentally enter the lungs, causing aspiration pneumonia. In many cases, it's safer to abstain from feeding the animal than to risk complications.

If a rehabber suggests administering rehabilitation fluids like unflavored Pedialyte, using small tools like a paintbrush for mice, an eye dropper for squirrels, or a tiny premature baby bottle for larger animals such as raccoons or fawns is recommended. Proceed cautiously and slowly. If you observe liquid exiting the nose, indicating it went down the wrong way into the lungs, it's critical to seek veterinary assistance immediately. Even with antibiotics, this situation is critical and may not have a positive outcome. Again, sometimes, providing no food or liquid is safer to avoid aspiration risk.

2. For baby animals entering rehabilitation, it's advisable to initially introduce them to rehydration fluids. After 24 hours, a gradual transition can be made to a solution comprising 25% formula and 75% water. This gradual adjustment helps acclimate their stomach to the new formula. A sudden switch from maternal milk to full-strength formula can shock their system and lead to various gastrointestinal issues. Never introduce a full-strength formula to an animal without proper hydration and gradient formula calculations.

3. Most animals, whether babies or adults in shock, require a heat source. Refer to the previously mentioned heat sources section for guidance. It's essential to provide a space for the animal to move away from the heat source when needed.

For an unconscious animal, the preferable approach is to place it on the lowest setting of a heat source. Ideally, incubators are the optimal choice, as they allow precise temperature control, ensuring the animal remains at the necessary warmth. Other methods are acceptable but require frequent monitoring.

4. It's crucial to emphasize that children and pets should never be allowed near wildlife. Doing so poses risks to the wildlife and the safety of children and pets.

5. Creating a warm, dark, and quiet environment is crucial—always warm, dark, and quiet! Loud noises like music, barking dogs, or excessive talking can significantly stress wild animals. Some creatures, such as raptors, cottontails, and white-tailed deer, are prone to capture myopathy. Treating these animals as you would want to be treated if an entirely unfamiliar being encountered you is important. Remember, these animals perceive you as abductors rather than helpers. Many are prey animals, so maintaining a calm, quiet demeanor is crucial. Avoid direct eye contact and sudden movements.

6. Remember, this isn't a pet or a child—it's a wild animal. It's essential to consistently respect its species and allow it to develop into the wild creature it's meant to be. Placing your desire for a pet over the animal's need for freedom is never fair!

7. If I can smell your perfume, cigarettes, scented candles, body odor, or air freshener, the animals can too. Avoid exposing them to these scents. Refrain from using deodorizers in your space; if you dislike the smell of the animals, reconsider rehabilitating animals and get them to your local rehabber quickly.

8. A healthy baby can easily go 24 hours without food. Don't rush a feeding because you don't hear back from a rehabber immediately.

9. Patience is vital. There are moments when there's nothing immediate to be done—sometimes, we can only wait.
10. First heat, second rehydrate, third feed. Animals must come up to heat before you hydrate. This takes time. Animals must fully rehydrate before they can feed. This also takes time, sometimes days. Don't rush. Trust the system.

Chapter 8
How to Proceed When Its an Emergency

Rescue Kit Items:

Being prepared to assist wildlife in times of need requires proactive preparation. Researching in advance and having accessible resources like this book can make a substantial difference.

Additionally, keeping a few essential items in your car or at your workplace can prove invaluable in aiding wildlife promptly. Consider storing a sturdy cardboard or plastic carrier equipped with the necessary tools:

- blankets
- towels
- thick gloves for personal protection
- latex gloves for handling
- a lightweight shovel (helpful in moving turtles or injured animals off roads)
- wire cutters
- scissors for traps or fences
- a portable heat source.
- For scenarios involving flying creatures, having a large butterfly or fishing net could aid in safely retrieving animals.

At home, maintain a collection of items suitable for various wildlife situations:

- appropriately sized cages or containers
- disinfectants, as outlined in this book
- petroleum jelly
- Pedialyte
- flea and tick treatments safe for kittens
- syringes
- eye droppers
- tweezers
- bandages,
- cotton balls
- bandage tape
- masks
- gloves

- towels
- blankets
- paper towels
- heavy protective gloves like welding gloves
- a can of puppy and kitten milk replacer, baby rice cereal and applesauce.

While it may seem like a collection of random items, these supplies can become crucial tools that save lives in urgent situations.

How to prepare a setup for an injured animal:

1. **Prep a Box:** Line a cardboard box with something absorbent, like an old t-shirt or hand towel.
2. **Glove up:** Do not touch the animal without gloves.
3. **Heat:** most animals will need a source of heat due to being too young or from being in shock.

Try to have your tools ready before moving forward. You don't want to waste time looking for your tools. When in doubt, give your local wildlife rehabber a call for advice. When rescuing an injured animal, it's essential to approach the situation carefully to ensure your safety and the animal's well-being:

1. Contact a Professional: If possible, contact a local wildlife rehabilitator or animal control center for guidance before attempting to rescue the animal yourself.
2. Assess the Situation: Evaluate the surroundings and assess the animal's condition from a safe distance. Approach slowly and calmly.
3. Protect Yourself: Use protective gear such as gloves or a blanket to prevent bites or scratches when handling the injured animal.
4. Handle with Care: Avoid sudden movements. If you must handle animals, do it gently. Injured animals are scared and may lash out.
5. Transport Safely: Use a sturdy, well-ventilated box or carrier to transport the animal, ensuring it has enough space and is secure during transit.

Remember, every animal regardless of size, has the potential to bite. A bite from a seemingly small animal, like a squirrel, can cause infection and pain similar to that from a larger animal. Additionally, identifying a snake as non-venomous doesn't guarantee safety—snakes too can deliver painful bites that may lead to infections. Raptors and songbirds possess beaks they'll use when feeling threatened, while the talons of Raptors can cause significant harm. Approach all wildlife with caution, using common sense to ensure your safety.

Understanding when to intervene and when to wait for help in wild animal encounters is crucial:

1. Animals in Need of Help: Identifying animals that need assistance is essential. Look for abnormal physical appearances, such as limping, multiple wounds, missing or swollen body parts, or abnormal behavior such as disorientation or aggression. These signs often indicate the need for human intervention. However, assessing whether the injury is severe or minor and whether it's recent or old is essential. Not all animals need rehabilitation; some may recover on their own. Consulting wildlife rehabbers or vets is crucial to determine the best course of action, considering factors like stress levels in rehab, seasonal implications, and the welfare of dependent young left behind.

2. Safety Concerns: There are instances where an animal may be in distress, but it's unsafe or inappropriate for you to assist. If you encounter an animal displaying severe neurological symptoms like

staggering, disorientation, or aggression, you must refrain from intervening and immediately contact local animal control or wildlife advocates. Ensure your pets are indoors and monitor the animal from a safe distance. Such behavior could indicate severe conditions like head trauma or potentially contagious diseases like rabies or distemper, which require professional expertise and protective equipment for handling. Organizations like A.E.R.O. provide training before deploying volunteers to aid. Don't risk yourself!

3. Risk Assessment: Assessing risk versus outcome is crucial. Animals struck by vehicles often face grim outcomes, and the likelihood of survival is low. While aiding recently injured animals might be beneficial, the risk of navigating high-traffic areas must be weighed against the animal's welfare. For slow-moving creatures like turtles crossing roads, signaling to slow traffic and aiding their passage can be immensely helpful if possible. However, it's crucial to prioritize human safety over wildlife assistance. Remember, risking personal safety for one animal might limit your capacity to help others. Ensuring your safety enables you to assist more animals effectively.

Once you've decided to aid an animal, ensuring its safety is priority #1. Avoid impractical or unsafe practices like tucking an animal into clothing for warmth. Instead, secure the animal in a suitable container with a heat source like a low-setting heat pad underneath a cardboard box or carrier.

Recognizing signs of a wildlife emergency often involves closely examining the situation. To gain a comprehensive understanding, wearing gloves and conducting a brief examination, if feasible, is crucial. Limit handling to under two minutes and minimize contact as much as possible. If feasible, capturing photos of injuries can also be beneficial. WARNING: Proceed with caution! If an animal is reactive or defensive, then wait. Never proceed if you're unsure or uncomfortable.

During an exam, always start with answering the big questions and take notes and photos when you can:

1. What is the nature of the situation? Identifying the animal's age and species is crucial. You can use the charts located in this book to help you determine this.
2. When did you first notice or encounter the animal? Prompt action is vital during emergencies
3. Where are you located? A rehabber may need you to provide the precise location where the animal was found. Beyond just your address, this specific location is vital for potential release if the animal can return home. Additionally, it offers crucial information for tracking factors like disease spread.
4. Who found the animal? You'll want to keep track of everyone who has handled the animal in case anything goes wrong.
5. How did the encounter happen? Understanding the circumstances of contact, such as a pet catching the animal or colliding with a windshield, helps assess the situation more accurately and provides better support once in rehabilitation.

Some obvious situations make an emergency for any animal. These include:

1. Maggots present on an animal
2. Bleeding or bruising/hematoma
3. Having been in another animal's mouth (pet cat, dog etc.)
4. Unconscious
5. An animal having been kept as a pet.
6. Approaching humans or inability to flee from humans.
7. A baby next to a dead adult
8. Hit by a machine (car, lawnmower, bike)

9. Odd head position
10. Hypothermia (too cold)
11. Hyperthermia (overheated)
12. Drooping wing
13. Stuck in a glue trap or netting.

Here are further details regarding these specific emergencies:

1. Infestation of maggots on animals: The presence of fly eggs or flies around an animal can rapidly escalate into an emergency. Fly eggs develop into maggots, which feed on flesh and cause irreparable damage if not promptly removed. In severe cases, euthanasia might be the best course of action.
2. Bleeding or bruising/hematoma: A hematoma, a swelling filled with blood, typically results from trauma. Any sign of blood, whether dried or fresh, indicates an urgent situation.
3. Injury from another animal's mouth: Any attack by a dog or cat should be treated as an emergency, even if initial damage seems minimal. Even small punctures from a cat can lead to severe infections. Dog attacks can cause hidden injuries, such as bone fractures, necessitating an X-ray.
4. Unconsciousness: An animal displaying closed eyes, a shallow pulse, and unresponsiveness indicates a critical situation.
5. An animal previously kept as a pet: Evaluation by a rehabilitator is necessary before deciding on release. Reptiles kept for over 30 days by the public are generally unsuitable for release by law.
6. Unusual behavior towards humans or inability to flee: This behavior, unless provoked by feeding from a member of the public, requires immediate attention.
7. Orphaned baby near a deceased adult: This situation signals that the baby is likely orphaned and needs assistance.
8. Injuries from being hit by a machine: Trauma from machinery can lead to significant issues. Getting these animals to a rehabilitation center as soon as possible is crucial.
9. Head tilt or abnormal positions: These signs could indicate various problems, from trauma to disease.
10. Hypothermia: Mammals and birds can suffer from being too cold. For reptiles, immediate access to a heat source is not crucial, as they are not affected by hypothermia due to being endothermic. However, they need a slow temperature increase to prevent shock.
11. Hyperthermia: Prolonged exposure to the sun can result in burns and overheating. Removal from the heat source and application of cool, not cold, moist towels to gradually lower body temperature is necessary avoiding sudden temperature changes that can cause shock.
12. Drooping wings in birds: An indicator of potential issues, often making capturing the bird easier.
13. Trapped in glue or netting: A trained rehabilitator should handle the removal to prevent further harm. Inexperienced attempts at removal can worsen the situation

More on infestations

Maggots, fly eggs, or fly strikes - a critical situation demanding immediate attention. Removing these creatures is not for the faint-hearted; but it's necessary. These are unhatched or, in the larval stage, insect larvae that feed on flesh. In warm weather, they can hatch within hours, leading to severe issues. They manifest as sawdust-like material or white/yellow clumps or granules.

It's crucial to thoroughly inspect inside the mouth and skin folds. To initiate removal:

- Use a flea comb soaked in diluted Dawn dish soap. This method proves effective in extracting them from fur.
- If the infestation seems overwhelming, applying cornstarch to the fur can initially dry up the eggs, making their removal easier once they are deceased.
- At times, shaving off the fur becomes necessary, but such action should never be taken without guidance from a wildlife rehabber. The fur plays a vital role in their survival, and its removal should be a thoughtful decision considering potential repercussions and exploring alternatives.

Handling this situation can be tricky if you notice maggots or fly eggs within a turtle's shell. Submerging responsive turtles in water can help eliminate live maggots.

In cases where eggs are spotted in the mouth or anus, a Capstar, a fast-acting flea treatment that animals ingest, can be mixed with water. If size-appropriate, it can be administered internally, or if especially small or baby animals, you can administer it externally by squirting directly on eggs or larvae. Mix ¼ tablet with 1cc warm water for oral administration, dropping it in the mouth one drop at a time. Alternatively, mix ½ tablet with 2cc of water following the instructions for external use, flushing the fly eggs/maggots from the body's exterior, paying close attention to the eyes, mouth, ears, and anus.

Fleas

Fleas, external parasites that feed on the blood of their host animal, are identifiable when you observe jumping or crawling movements and/or notice brown or red flecks on the skin. These pests can also transmit tapeworms and potentially cause severe anemia, particularly in small or young animals.

To address flea infestations:

Employ a flea comb soaked in diluted Dawn dish soap to remove them from the fur effectively. If the animal can be safely handled, consider using a warm diluted solution for bathing, although this should ideally be carried out by a professional or under the guidance of a rehabber. Kitten-safe pyrethrin-based flea products like Revolution can be utilized in small doses after a dawn bath. Another effective option is oral Capstar. Prepare a mixture of ¼ tablet with 1cc warm water for oral administration, administering it one drop at a time.

Botflies

Botflies lay eggs that hatch and burrow under the host animal's skin. Initially, they resemble pimple-like bumps with tiny breathing holes. These parasites grow into marble-sized masses and eventually exit through these holes. While the host may survive, cleaning the wound is essential if possible. Particularly on small mammals, botfly larvae can be devastating, depending on their location.

Removing botflies is best left to highly trained individuals capable of sedating the animal. Veterinarians might surgically extract the larvae, ensuring complete removal to prevent potential complications. Attempting removal without expertise might result in leaving parts of the botfly behind, leading to more significant issues. The larva's spine can induce pain. Hence, attempting at-home removal isn't recommended.

A recommended approach for small, early-stage bumps is applying petroleum jelly. Once the fly dies, consulting with a rehabber about larva removal or allowing natural rejection is advisable.

In advanced cases, it's prudent to seek a vet's assistance or let the botfly naturally eject itself. Typically, the botfly's natural cycle doesn't fatally harm its host. Although uncomfortable and visually unpleasant, adult squirrels often experience botflies without severe side effects. However, each case varies, so individual assessment is crucial.

Ticks

Ticks, static parasites from the arachnid family, feed on the host's blood externally. They appear as swollen bumps ranging from the size of a pinpoint to that of a pea, contingent on their engorgement. These parasites attach to the host using their mouthparts.

Commonly found on cottontails or white-tailed deer, ticks can harbor diseases transmissible to their hosts.

For ticks not profoundly embedded, removal can be done using a comb, dropping them into a cup of alcohol. If the tick's head is embedded, tweezers can be used. Firmly grasp around the head, pulling straight out, away from the skin. Ideally, the head will be extracted cleanly. If not, stay calm and monitor the spot for signs of trouble in the future. Swelling, discharge, or persistent redness are potential issues requiring a vet or rehabber consultation.

Chapter 9
Renesting and Reuniting

Reuniting or renesting is suitable only for healthy, uninjured babies. Understanding the baby's age and developmental stage is crucial in determining its life phase. It requires substantial experience to gauge a baby's age and understand its specific needs. To assist with this, we're creating charts for each species. These resources will allow you to visually assess and identify the animal's age, aiding in making the most appropriate decisions for their care.

Birds

Understanding the stages of a bird's development is crucial to the renesting process. Baby birds progress through three significant developmental stages: hatchling, nestling, and fledgling.

<u>Hatchling:</u> These are naked with closed eyes. Sometimes, they have some fluff, but their eyes will remain closed.

An example of hatchling babies

Nestling: They have pin feathers down and can open their eyes. They'll still close their eyes when they sleep.

This is an example of nestling babies

For a hatchling or nestling, searching for a nearby nest is vital. Observe the nest's description and the baby's appearance. Scan trees, gutters, vents, and nearby structures within a 20-foot radius of where the baby was found. Remember, they can't keep themselves warm at this stage, so that heat may be needed.

- If an adult is nearby and appears attentive, try renesting.
- If a nest is found, return the baby. If it's cold, warm it for 30 minutes first. Ensure consistency in appearance with other nestlings.
- If no nest is found, create an artificial nest and hang it nearby. Maintain a distance to allow the mother to return 1.5 hours for nestlings and fledglings. Not an option for hatchlings as mom won't be able to keep two nests warm.
- There are myths about renesting baby birds, like the belief that a mother bird won't accept her young if they smell like humans. This is untrue. Mom doesn't mind a little human assistance if you do it correctly.

Fledglings: Covered primarily on feathers with a short tail, they are learning to fly and may have left the nest. You may often see fluffy, downy feathers still sticking out. Parents continue caring for them; giving them time and space typically allows them to fly without issues.

- Fledglings are often spotted hopping around on the ground. The parents continue to care for and feed these babies until they're old enough to fend for themselves. If you find a fledgling in a dangerous spot, like the middle of a road, carefully pick it up and move it to a safer nearby area. Never relocate the bird!
- Sometimes, concerns arise about outdoor cats posing a threat to these birds. It's a valid concern. The best solution during the fledgling season is to keep the cats indoors for a few days. Remember, cats aren't native, and allowing these birds to live their natural lives is essential.

- If someone you know brings home a fledgling bird, return it to where it came from as quickly as possible! I've had parents accept fledgling birds back two days after they were brought in. It's always worth a try!

This is an example of a fledgling bird

Renesting different types of Birds:

Recognizing the bird species is crucial in providing appropriate assistance. If you lack extensive experience, seeking guidance from a professional for precise identification is advisable. Once you ascertain the species, you can reference pictures of their nests to determine if they are cavity, cup, or tree nesters, aiding in their care.

- Cavity nesters tape shut a cardboard box, cut an entrance hole, and add natural material. Hang it near where the bird was found, avoiding direct sunlight.
- Cup and tree nesters can use plastic plant pots with drainage holes, also filled with natural material, hung in shaded areas.

Once you've returned them home, keep an eye out for parents. There's nothing more rewarding than seeing Mom and Dad come back to feed them!

In cases where a baby animal is unable to stand or move properly, shows signs of injury like wounds or bleeding, has feathers soaked in water or any sticky substance, or if there's evidence of it being in another animal's mouth, even without apparent physical injury, it's advisable to seek the help of a rehabber.

Mammals

1. Put on gloves immediately and move the baby into a cardboard box with a blanket and a heat source.
2. Start trying to identify the baby.
3. Check the area for other babies. If multiple are found on the ground, it might suggest nest destruction or the mother's absence.

Examples of reunification setups:

When attempting to reunite, a cardboard box serves as an excellent temporary setup. Sometimes, babies can fall from trees or be left behind if their mother becomes startled or when they get adventurous and explore early. Placing them in a cardboard box with soft bedding and a heat source is ideal. Using cardboard also allows for leaving a note in public areas, indicating that wildlife rescue has been contacted and reunification is ongoing. Placing the box where adult animals frequent often results in the adult finding the baby and returning it to its home. In cases like squirrels, playing a sound clip of a squirrel baby crying can attract the mother's attention and prompt her to return and care for the baby.

Heat Sources

The most reliable heat source for this purpose is an incubator. Numerous brands are available, and many rehabbers rely on Brinsea models. Babywarm.org organizes fundraisers specifically for these incubators. If you're keen on fundraising for an incubator, please contact your sponsor for more details.

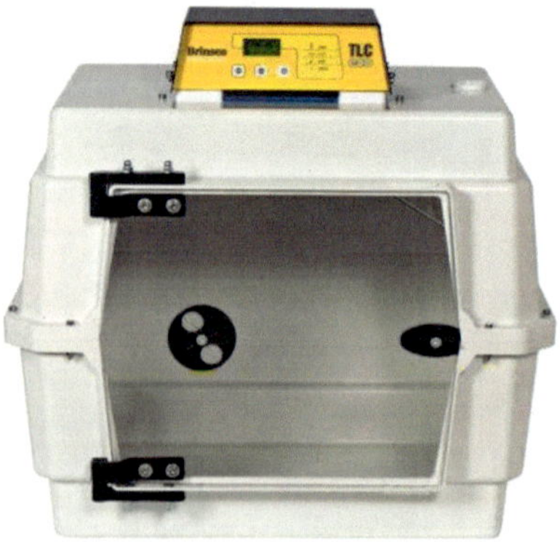

Heat discs are also an excellent heat source. After microwaving them for about 4 minutes, they can retain heat for up to 11 hours, making them ideal for car trips! Ensure the cover has Velcro running across to prevent babies from climbing inside.

This is a good setup with a heat disc suitable for most small mammals.

Heat pads are fantastic but exercise caution. Some models have an automatic shut-off feature after a few hours. Always ensure the heat pad is set to a low temperature for safety.

Heat socks are a handy option in a pinch and can even be a fun project for kids to decorate. Just a note of caution: if using non-permanent markers, be aware that the colors may come off when heated in the microwave. Heat the sock for 30 seconds initially and test the warmth. Once warmed, wrap it in a hand towel as the heated rice inside produces steam heat. Caution is advised to avoid burns!

Contain It

Let's break it down based on species, helping you understand what to include in the box to ensure the animal's safety.

Squirrels/rats/mice: Refer to our included age charts for guidance. Pinkies, for instance, can be placed in a makeshift nest positioned at the base of a nearby tree while playing a YouTube sound clip of a squirrel/rat/mouse crying. Typically, these clips are about an hour long. Allow space and observe for the mother. It's best to avoid admitting pinkies to rehab unless all reunification attempts fail, as many mothers retrieve them. These young ones demand continuous care, consuming considerable time for local rehabbers. If you must take them to a rehabilitator, inquire about their needs; sometimes, providing fresh veggies and paper towels can be immensely helpful.

Groundhog: Search for a den site and give the mother five hours before considering other options. Look for signs such as dirt piles or holes in the ground. Avoid leaving the animal out overnight. It's recommended to discourage these animals from entering rehab unless all attempts at reunification fail.

Opossums: Baby opossums typically stay with their mother, either in her marsupial pouch or clinging to her back. If a baby less than 8 inches is found alone, it should be taken into rehab, as reunification often doesn't succeed unless a nearby mother is confirmed.

Foxes: Both fox parents usually care for their babies. As foxes are nocturnal, leaving them out overnight is suitable. Ensure a heat source is with the baby, placing it in the path frequently used by the fox. Clearly label the box for awareness. It's advisable to avoid admitting foxes to rehab unless all reunification efforts fail.

Bats: Young bats might fall from colonies. Attempt to place the baby on a tall ladder, broom handle or sticks, guiding it back to the colony entrance. It should crawl back in. If unsuccessful or unable to locate the colony, bring the baby in promptly.

Skunks and Raccoons: These nocturnal animals should be left out overnight, ensuring a heat source is available with the baby. Place them in the animal's usual path, with clear labeling to notify others of ongoing rescue efforts.

Cottontails: Nests of baby cottontails are often disrupted. If the nest spot can be located, reconstruct it; otherwise, these babies should be brought into rehab, as mothers don't typically search for displaced or relocated babies due to stress susceptibility. Avoid excessive handling.

If uncertain whether the mother will return to the nest, you can attempt a simple technique. Place sticks in a cross shape over the nest or create a circle of flour around it. If the sticks are disturbed or the flour circle is disrupted, the mother has visited the nest.

Reptiles

Fortunately, orphaned reptiles are a rarity, which is a relief! If you come across uninjured baby reptiles, the best course of action is to return them outdoors to where they were found. However, if an animal is stuck or has fallen into a potentially dangerous situation like a window well, it's important to gently remove them from that predicament. After rescuing them, it's best to relocate them very close to their original location but in a safer environment, away from harm.

Chapter 10
Normal vs Abnormal

Cottontails

Normal:

- Active during dawn and dusk (crepuscular).
- Independent at 3-4 weeks old.
- Diet includes greens.
- Raised by a single mother with no parental assistance from the father.
- Lifespan averages around 2 years.
- Exhibits flight, fright, or fight responses are typical behaviors.
- Babies are born between February and September; mothers can have up to 6 litters yearly.

Abnormal Signs:

- Eyes-closed baby found outside the nest.
- Head tilt observed.
- If found in an animal's mouth, immediate rehab is necessary.

Preparing for Rehab:

- Ensure a calm, quiet, dark, and warm environment, as rabbits are susceptible to stress.
- Place the rabbit in a box with soft bedding. If necessary, bring the entire fur-lined nest.
- Eyes-closed babies or injured adults need a heat source.
- Use gloves due to the risk of carrying zoonotic diseases. Ensure that the finder's pets are on a monthly preventative dewormer.

Newborn-1 week.

Less than 60 grams

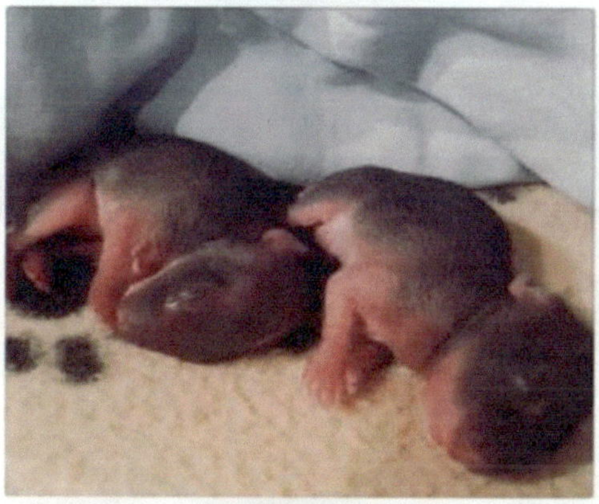

Newborn to one-week-old cottontail rabbits are characterized by closed eyes and ears that lay flat against their backs. Resembling miniature baby hippos, they possess a distinctive appearance with a rounded, slightly bloated body. Their fur might appear sparse, revealing a purplish hue, contributing to a bloated or swollen appearance, indicative of their early developmental stage. This phase marks a time of dependency on their mothers or nesting sites, where they rely on warmth and nurturing for survival.

1-2 weeks

60-80 grams

Around 1 to 2 weeks old, cottontail rabbits experience changes in their appearance and development. Their eyes remain closed, but their ears shift, transitioning from the earlier laid-flat position. Fur starts to emerge across their bodies, gradually covering their once-bare skin. At this stage, they no longer resemble the distinctive appearance of a baby hippo but instead, exhibit the early signs of a developing rabbit.

2-3 Weeks

80-100 grams

Between 2 to 3 weeks old, cottontail rabbits undergo significant developmental changes. Their eyes open for the first time, giving them an initial view of their surroundings. Their ears remain backward, and their fur grows, covering more of their bodies. At this stage, they exhibit increased activity, displaying behaviors like "popcorning," where they make sudden, quick hops or jumps in excitement or reaction to stimuli. They become more responsive and reactive to their environment as they explore and engage with the world around them.

4 weeks and older

100+ grams

By 4 weeks and beyond, cottontail rabbits show substantial development. Their eyes are fully open, allowing them to keenly observe their environment. Their ears are upright, erecting and becoming more prominent as they transition towards looking like miniature versions of adult rabbits. Their responsiveness to stimuli and surroundings heightens significantly, displaying heightened reactivity and sensitivity. This period can also induce high stress as they adapt to their environment and develop survival instincts.

Squirrels

Normal Squirrel Behavior:

- Diurnal (active during the day).
- Babies become independent around 2-3 months old.
- Diet includes greens, vegetables, fruit, seeds, and nuts; rarely carnivorous.
- Raised by a single mother, the male defends the territory and brings food but may harm the young.
- Wild lifespan averages about 6 years.
- There are two yearly baby waves: February through April and September through November.
- The average litter size is 2-3.
- A nest is a high ball of leaves in a tree, typically with 2 nest sites.
- Vocalize loudly when disturbed.
- All squirrels have black fingernails.
- Pinkies should have a milk line, as shown in the picture.

Abnormal Squirrel Behavior:

- Eyes-closed baby found out of the nest.
- Young juveniles approaching humans.
- Head tilt observed.
- If found in an animal's mouth or experiencing a fly strike, immediate rehab is needed.

Preparation for Rehab:

- Ensure a calm, quiet, dark, and warm environment.
- Attempt reunification but bring babies in after dark if unsuccessful, resuming efforts in the morning.
- Place in a box with soft bedding, or bring the whole nest if necessary.
- Eyes-closed babies or injured adults should have a heat source.
- Use gloves due to potential zoonotic diseases and ensure the finder's pets are on a monthly dewormer.
- Do not feed any formula without guidance from a rehabber.

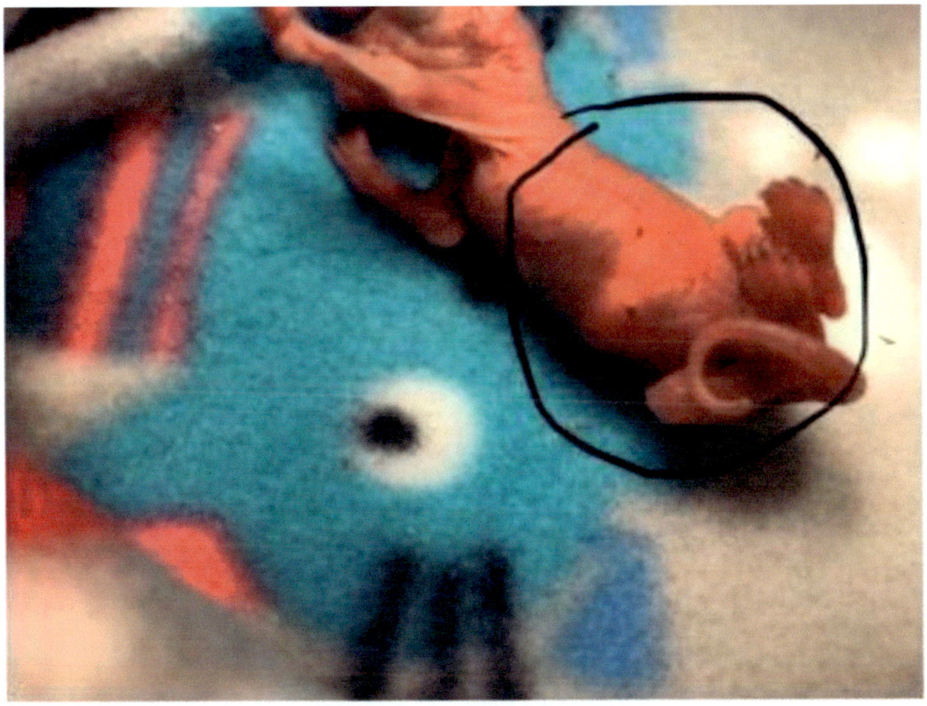

In this picture, you can see that the babies are born without fur, and their eyes are closed. If you look closely, you'll also see that the babies always have black fingernails, which helps to ID them. Additionally, you'll be able to tell from looking at the belly that there is a white portion in the picture below. This is referred to as a "Milk line," a good sign that the mom has tended to the babies recently.

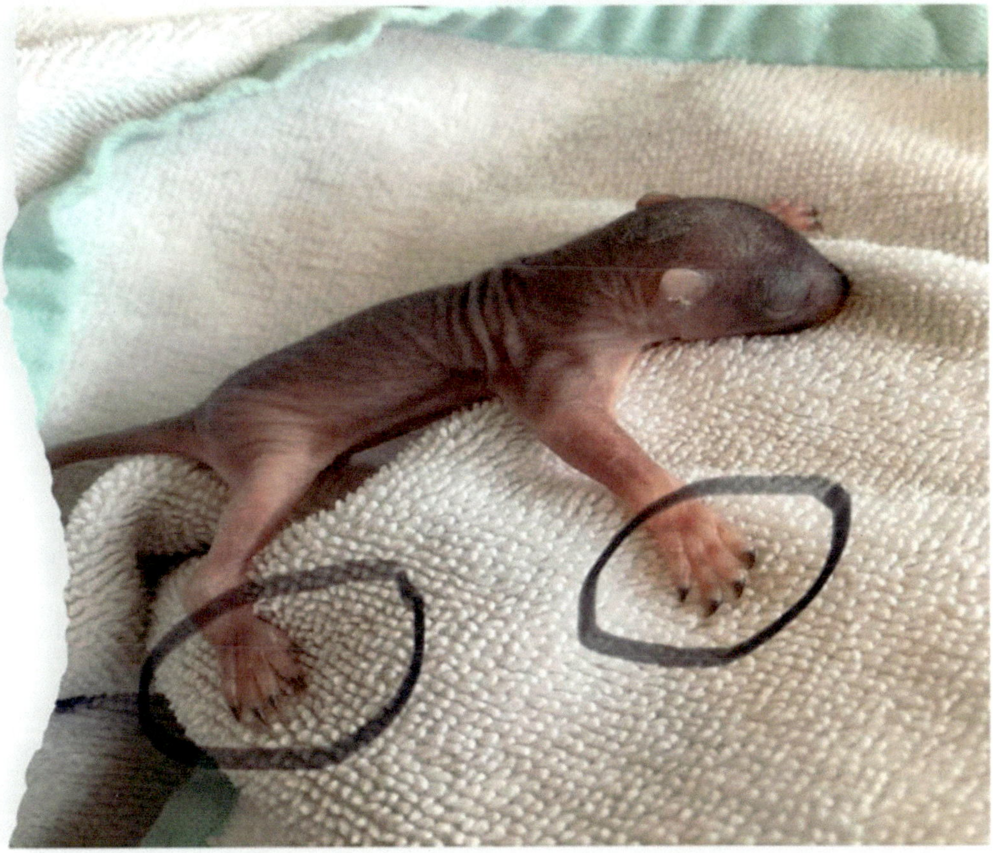

0-1 week,

10-30 grams

KEEP WARM!

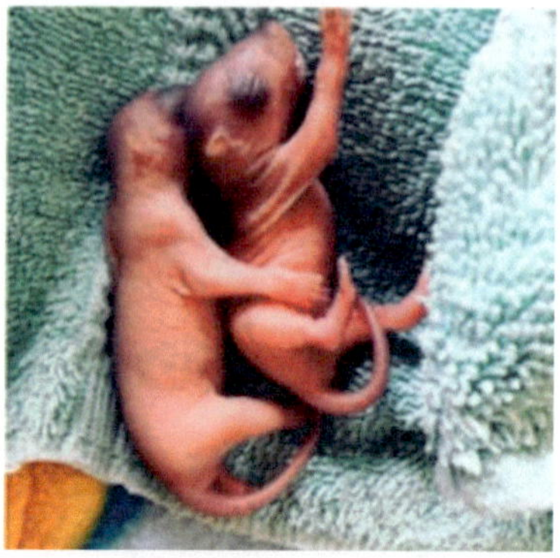

Pink, no pigment, eyes and ears fused shut.

Underdeveloped Looking. This age needs to be fed every 2-3 hours.

1-2 weeks

30-45 grams

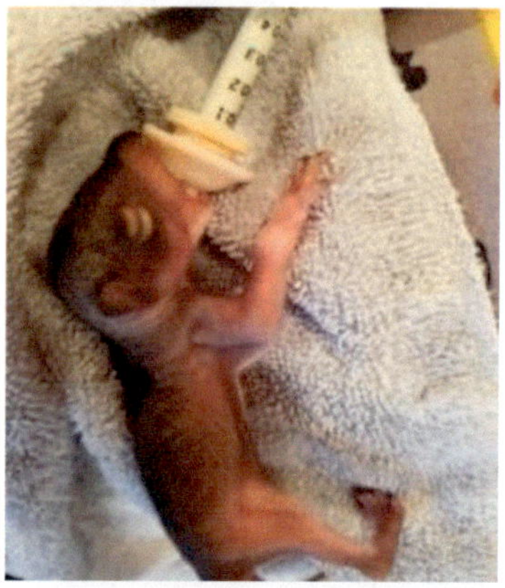

Pigmented, eyes still closed, starting to walk on its own but unstable

2-3 weeks

45-75 grams

Fur grown, ears up, very mobile, climbing around the cage floor.

Continue with heat. These babies are now eating every four hours.

3-4 weeks

75-100 grams

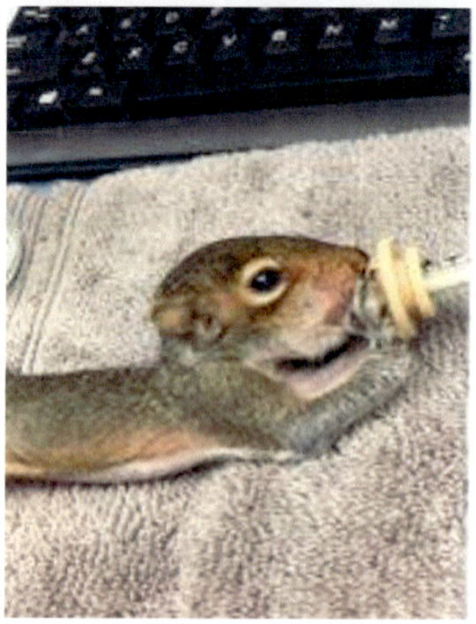

Full coat of thin fur, markings coming in,

5-6 weeks

100-120 grams

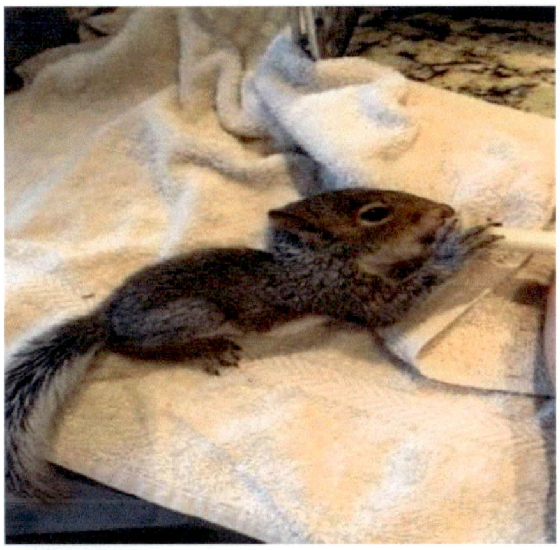

Eyes open, climbing, vocal, tail starting to fluff, and can start to come off heat.

Bottle feeding every 4 hours and introducing solid foods.

7-8 weeks

120-150 grams

Able to sit on hind legs, eating more solids, beginning to look like more of an adult squirrel than a pinkie.

9-12 weeks

150-180 grams

Looks like a big kid squirrel, but it's still taking a bottle once a day. Don't be alarmed if they completely wean themselves; follow their lead.

These are now at a weaning stage.

12 weeks and older 180+

BIG KID! You should no longer be able to handle it. Weaned. Can crack a nut on its own.

Bats (this information is based on the Big Brown Bat, which is a prevalent species in VA)

Normal Bat Behavior:

- Nocturnal, active during the night.
- Babies become independent at around 1 month old.
- Most North American Bat diets consist of insects (insectivores).
- Raised by a single mother, often lives in maternal colonies with other females and babies, while bachelor colonies stay separate.
- Wild lifespan averages about 10 years but can extend up to 30 years
- Most babies are typically born at the beginning of June-July.
- The average litter size is 2; the mother usually has one set of twins.
- Most species have colonies gathered to roost; roost sites vary based on species.
- Bats have teeth and can bite—handle with care!

Abnormal Bat Behavior:

- A naked baby bat was found alone.
- The bat was found on the ground, unable to fly.
- Bats found inside homes.
- Bats that have bitten must be trapped and tested for rabies immediately. Contact the local health department.

Preparation for Rehab:

- Ensure a calm, quiet, dark, and warm environment.
- Provide soft bedding in a box. Babies appreciate a stuffed animal, while adults prefer something to hang from.
- Eyes-closed babies or injured adults should have a heat source.
- Use gloves—garden gloves with latex gloves underneath are advisable.
- Move slowly as their hook-shaped nails can cling to surfaces; avoid pulling on their toes.
- Do not feed any formula without guidance from a rehabber.

Picture 1: you'll notice the baby bat has a white spot that you can see through the translucent skin. This is a milk line. It can show that the baby bat does have a belly full of milk. This baby should often try to be reunited with the colony.

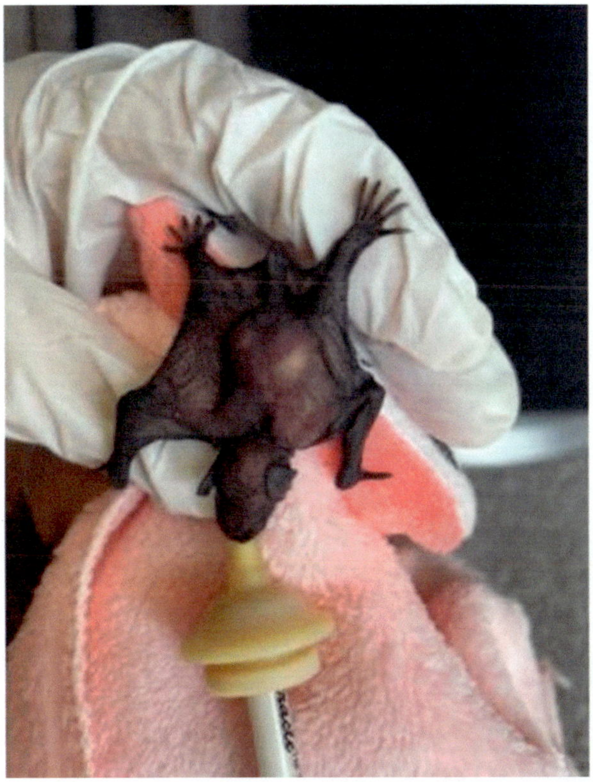

Picture 2: You can see the jaws in picture 2. Their bites can break the skin. Handle with caution!

In the chart below, you'll see the standard weights for the most common bat in rehab here in VA. The big brown bat. Keep in mind these weights are a little high for most other species. Try your best to Identify your species before assuming anything.

Newborn

Under 5 grams

Baby bats, often referred to as pups, are born hairless, resembling tiny, winged gargoyles. With closed eyes, they rely on their ears to navigate their surroundings and typically cling to their mothers or the roosting site for safety and warmth until they mature.

2-5 days

5-8 grams

*adults of many North American species can be full grown at 5-8grams

Around 2 to 5 days old, baby bats begin to develop a fine peach fuzz on their previously bald bodies. Their eyes also open during this period, allowing them to perceive their environment. At this stage, they gain some mobility and can begin crawling, displaying early signs of movement and exploration within their surroundings.

1-2 weeks

8-15 grams

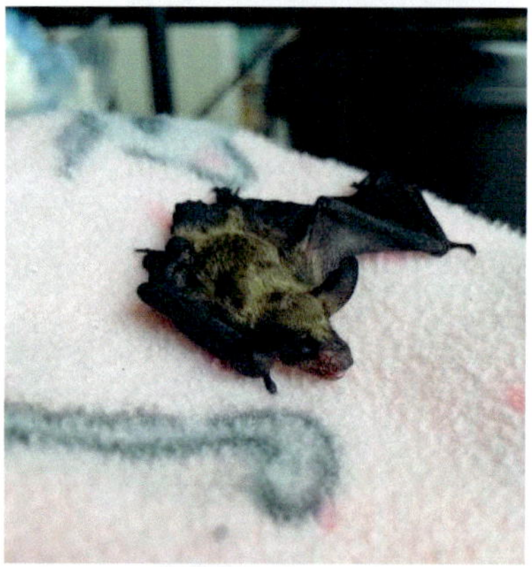

Around 1 to 2 weeks old, baby bats become fully furred, with their bodies covered in a coat of fur, resembling miniature versions of adult bats. During this time, they start practicing flapping motions with their wings, preparing for eventual flight. It's important to note that at this stage, they may exhibit defensive behavior and bite if they feel threatened or scared.

3-4 weeks

16-22 grams

Baby bats remain fully furred around 3 to 4 weeks old and require ample space and exposure to the outdoors to develop their hunting and flying abilities. This period is critical for their development as they practice flying maneuvers, hunt for food, and refine their skills in navigating their natural environment.

Whitetail Deer

Normal Deer Behavior:

- Crepuscular, active during dawn and dusk.
- Fawns born in May-June in VA, with 1-2 babies per litter.
- Highly social, living in herds.
- Skittish prey animals.
- Fawns curl up in a ball while sleeping.
- Fawns have no scent for the first few days of life.
- Diet includes greens, vegetables, fruits, and available vegetation.
- Fawns are born with eyes open and fully furred.
- Mothers place babies in one spot, like under a bush, and visit at dawn and dusk to feed.
- Parents remain nearby; avoid assuming the baby is abandoned.
- Fawns lose their spots at around 1 year.
- Proficient swimmers, fast runners, and capable of clearing an 8.5-foot fence.

Abnormal Deer Behavior:

- Fawn wandering, calling out for over an hour.
- Presence of flies, maggots, or ants.
- Laying down laterally.
- Dead doe nearby.
- Assisting adults can be challenging and hazardous! It may also be illegal, depending on your state laws.

Preparing for Rehab:

- Ensure a calm, quiet, dark, and warm environment.
- Place the baby in a laundry basket or playpen with soft bedding.
- Play the sound of a crying baby loudly and observe for parents (they may come quickly but remain hidden).
- Baby should have a heat source or be kept warm, avoiding cold environments.
- Move slowly to minimize noise and reduce stress.
- Do not feed any formula without permission and guidance from a rehabber.

You'll notice many different sizes and age stages in this picture. The second from the left is the smallest and still has its spots and darker rust color. The tallest one is still young but no longer has spots and is turning a lighter fur shade.

Foxes

Normal Fox Behavior:

- Nocturnal, active at night.
- Live in social units but are not inclined to stay together long term.
- Carnivorous diet—consumes small mammals, birds, and cat food left out.
- Both parents raise babies; the father feeds the mother while she stays with the babies for 2-3 weeks, only moving if the den is disturbed. If the mother is absent, other family members step in to maintain the feeding schedule, so avoid quick assumptions of abandonment.
- A wild lifespan is approximately 5 years, but it can be up to 15 years in captivity.
- Share territories within their family units but do not often tolerate outsiders.
- Crucial to release them back where they originated.
- Young foxes might approach humans as they initially perceive no threat, but this changes as humans and pets claim territory.

Abnormal Fox Behavior:

- Baby foxes wander alone, crying excessively potentially becoming separated, especially at 6 weeks+ when they become more mobile.
- Presence of flies, maggots, or ants.
- Laying down laterally.
- Dead adult(s) nearby.

Preparing for Rehab:

- Maintain a calm, quiet, dark, and warm environment.
- Babies require heat and hydration.

- Do not feed any formula without guidance from a rehabber.
- Use gloves; latex gloves paired with heavy "cat gloves" are recommended.
- Keep separated from other pets and children.
- Assisting adults can be difficult and hazardous. Utilize humane traps for adults for treatment; avoid handling without experience.

This is a picture of a fox with a mange mite infestation.

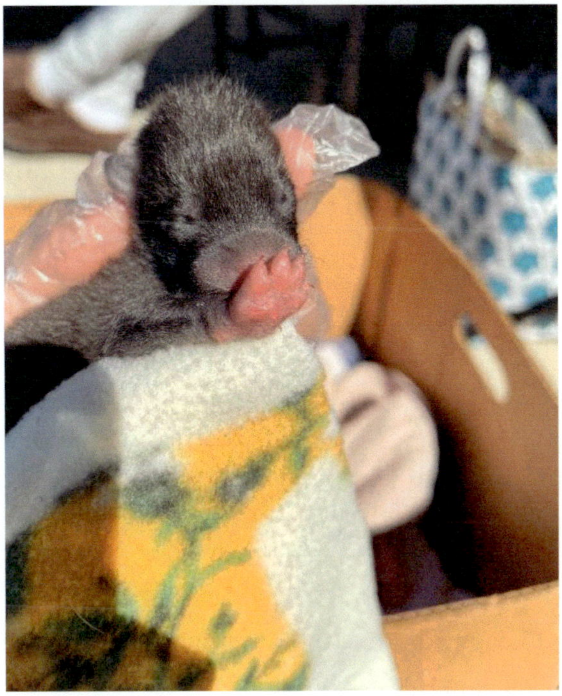

This is a picture of a baby fox. They change quite a bit throughout growing. They’re born furless dark-skinned.

This photo is another excellent example of two different ages of Red Foxes. The younger red fox without black markings and the more "Teenage" red fox with markings are beginning to look very adult.

Groundhog

Normal Groundhog Behavior:

- Diurnal, active during the day.
- Makes barking or whistling sounds.
- Proficient swimmers and climbers; adept at tree-climbing to avoid predators.
- Live in colonies but space out their dens.
- Relies on greens and vegetation for survival.
- They will consume fruits and veggies when available and might eat small amphibians or baby birds found on the ground.

Abnormal Groundhog Behavior:

- Baby groundhogs wander alone, crying excessively, potentially becoming separated from the family, especially after 6 weeks when they become more mobile.
- Presence of flies, maggots, or ants.
- Laying down laterally.
- Dead adult(s) nearby.

Preparing for Rehab:

- Maintain a calm, quiet, dark, and warm environment.
- Provide heat and keep babies hydrated.
- Refrain from feeding any formula without permission and guidance from a rehabber.
- Wear latex gloves, and heavy "cat gloves" are recommended.
- Keep separated from other pets and children.
- Assisting adults can be difficult and hazardous. Utilize humane traps for adults for treatment; avoid handling without experience.

These babies are quite small but very much resemble their parents. You can see the powerful hands for digging.

Chipmunks

Normal chipmunk Behavior:

- Diurnal, active during the day.
- Babies are independent at approximately 2 months.
- Diet includes greens, vegetables, fruit, seeds, nuts, buds, occasional insects, worms, frogs, baby birds, grass, shoots, and mushrooms. May consume items from gardens.
- Only the mom's parents; the male defends the territory and brings food for the mother. However, the male may sometimes kill young.
- Wild lifespan is about 3 years.
- Two waves of babies each year.
- Litter size is typically 4-6.
- Constructs burrows underground with multiple entrances.
- Can sleep up to 15 hours a day.
- Makes chirping noises.

Abnormal chipmunk Behavior:

- Eyes closed, baby out of the nest.
- Young juvenile approaching humans.
- Head tilt.
- MUST SEEK REHAB IF IN AN ANIMAL'S MOUTH.
- Fly strike.

Preparing for Rehab:

- Maintain a calm, quiet, dark, and warm environment.
- Attempt to reunite but bring babies in after dark. Continue attempts in the morning.
- Place in a box with soft bedding and keep it dark.
- Ensure that any eyes closed to babies or injured adults have a heat source.
- Wear gloves due to potential zoonotic diseases. Ensure the finder's pets are on a monthly preventative dewormer.
- Do not feed any formula without permission and guidance from a rehabber.

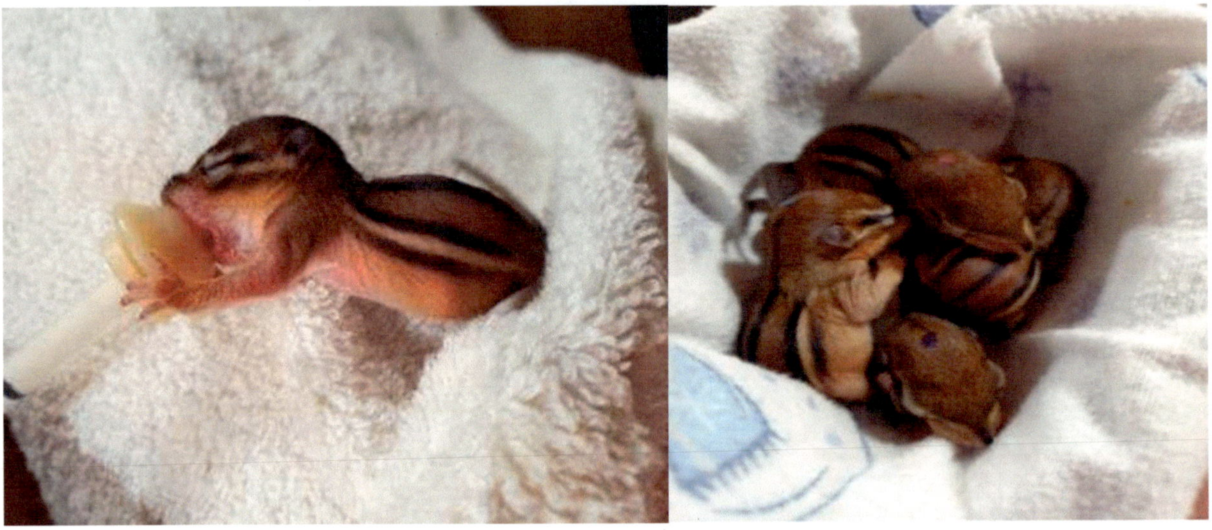

These baby chipmunks already have their signature stripes down their backs, even when young.

Opossum

Normal Opossum Behavior:

- Nocturnal, active at night, limited eyesight during the day.
- Only North American marsupials. Babies reside in mom’s pouch or on her back until 15-16 weeks.
- Solitary animals, though babies, stick close to mom until they're ready to leave. If found alone and not of appropriate age, it likely needs rehab.
- Omnivorous, consuming a wide range of food, including carrion and, in captivity, other opossums.
- A wild lifespan is around 2-3 years, up to 8 years in captivity.
- Most active in Summer, Spring, and Fall; they don't hibernate but reduce activity in winter to conserve energy for breeding. Moms can have up to 3 litters annually.
- Born the size of a jellybean, permanently attached to mom’s tube-like nipple until 8 weeks. The low success rate for babies under 20g. Refer to picture.
- He opens his mouth to make a scary face as an average defensive behavior.

Abnormal Opossum Behavior:

- Baby wandering alone; any opossum less than 8 inches is considered a juvenile.
- Presence of flies, maggots, or ants.
- Laying laterally (on its side).
- Dead mother nearby.
- Head tilt.

Preparing for Rehab:

- Maintain a calm, quiet, dark, and warm environment.
- Ensure babies have heat and stay hydrated. The babies need high humidity.
- Do not feed any formula without a rehabber's permission and guidance.
- Wear gloves, preferably latex, with heavy-duty "cat gloves." or a thick towel.
- Keep separate from other pets and children.

0-3 weeks,

10-25grams

Only attempt to rehab 15+

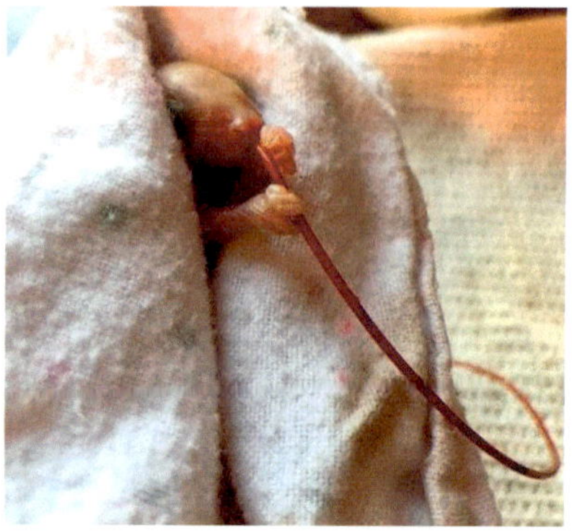

Newborn opossums display a distinctive appearance from birth to three weeks old. They're characterized by their pinkish hue, lacking pigment, and appear underdeveloped compared to many other mammalian newborns. Their eyes and ears are fused shut during this stage, contributing to their overall underdeveloped look.

Being marsupials, opossum babies are born undeveloped compared to many other mammals. Their early development occurs externally as they grow and mature within their mother's pouch. The pouch provides a nurturing and protective environment where these tiny creatures undergo the crucial stages of development until they become more self-sufficient.

4-6 weeks

25-45grams

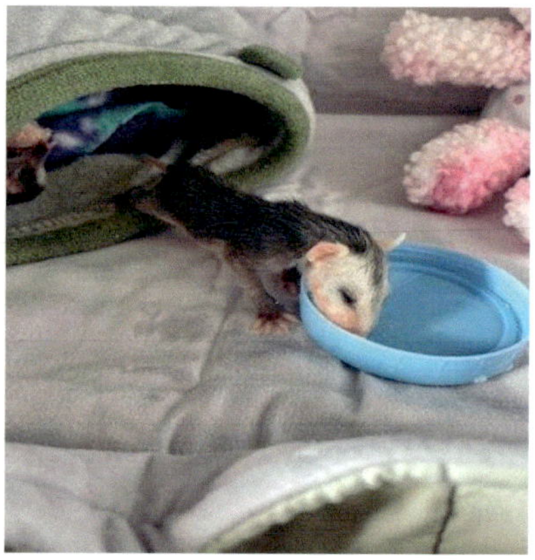

Between 4 to 6 weeks old, opossums exhibit noticeable changes in their appearance and abilities. Their pinkish hue transitions to a more pigmented coloration as their bodies develop pigmentations. Despite this, their eyes remain closed during this stage of development. They begin to demonstrate early signs of mobility, attempting to walk independently, but their movements are often unsteady and unstable as they explore their surroundings. This period marks the initial attempts at self-propelled movement, although they still heavily rely on their mother's care and guidance for survival.

9-11 weeks

60-100grams

Between 9 to 11 weeks old, opossums undergo significant development. Their fur has fully grown at this stage, providing them with a more complete and mature appearance. Their eyes have opened, revealing their surroundings for the first time, and their ears are upright, enhancing their ability to perceive sounds in their environment. They are highly mobile, displaying increased movement and agility, though they might not have mastered climbing yet. This period marks a phase of increased independence and exploration as they gradually acquire more skills to navigate their surroundings.

11-15 weeks

100-200g

Between 11 to 15 weeks old, opossums exhibit further growth. They'll have a fluffy, full fur coat, with distinct markings beginning to emerge, contributing to their unique appearance. During this period, they often display skittish and fearful behavior, being easily startled or jumpy in response to stimuli in their environment.

At this stage, providing plenty of food in multiple dishes for these opossums is beneficial. Offering a variety of nourishment in different dishes ensures they have access to sufficient sustenance, supporting their continued growth and development during this crucial phase of their lives.

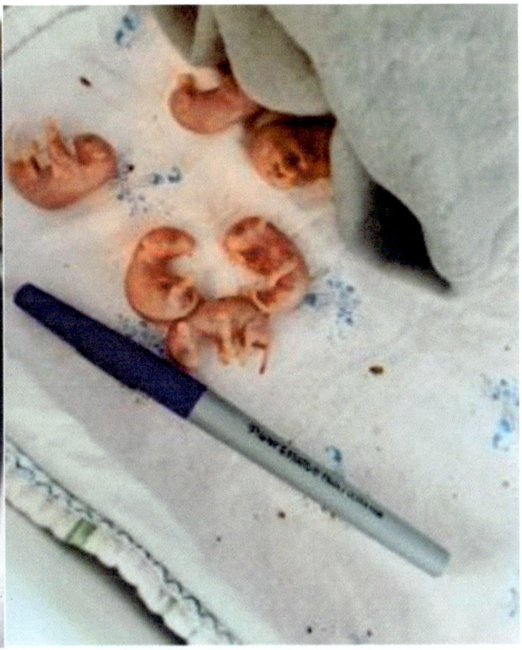

The aging process of opossums is remarkably unique, making it crucial to comprehend the full range of ages. Examining these two pictures shows that significant differences exist between newborn opossums and those just a few weeks old.

Raccoon

Normal Raccoon Behavior:

- Nocturnal, primarily active at night. Nursing females might emerge during the day for extra food.
- A carnivorous diet includes small mammals, birds, and occasionally cat food left out.
- Only the mother raises offspring; males may harm babies.
- Proficient swimmers and climbers often den near water.
- A wild's lifespan is around 2-3 years, but they can live up to 20 years in captivity.

- May share territories with external family units but are intolerant of outsiders.
- Possess a susceptible sense of touch; tend to grab and manipulate objects.
- Known for intelligence, though, can exhibit challenging behavior.
- Babies are born from April to June, with 2-4 per litter. They remain dependent on their mom for 3 months.

Abnormal Raccoon Behavior:

- Wandering baby crying excessively, especially if alone and starving.
- Presence of flies, maggots, or ants.
- Laying laterally.
- Dead adult(s) nearby.
- Hairless appearance.

Preparing for Rehab:

- Create a calm, quiet, dark, and warm environment.
- Ensure babies have heat and stay hydrated.
- Do not feed any formula without permission and guidance from a rehabber.
- Use gloves; latex gloves with heavy-duty "cat gloves" are recommended.
- Keep separate from other pets and children.
- Attempting to assist adults can be challenging and hazardous.
- Utilize humane traps for adults for treatment; avoid handling without experience.

Skunk

Normal Skunk Behavior:

- Nocturnal, primarily active at night.
- Poor eyesight but possesses a potent defense mechanism.

- Diet mainly consists of insects, vegetation, and occasionally cat food.
- Defensive behavior can seem fearless until something gets too close.
- Dens in various locations, often near water.
- Lifespan averages around 2-3 years in the wild but can live up to 15 years in captivity.
- Babies are typically born from April to June, with an average litter size of 5-6.
- Cannot spray until they are 4+ weeks old.

Abnormal Skunk Behavior:

- A baby under 7 inches wandering alone and crying excessively may approach humans if alone and starving.
- Presence of flies, maggots, or ants.
- Laying laterally.
- Dead adult(s) nearby.

Preparing for Rehab:

- Ensure a calm, quiet, dark, and warm environment.
- Provide heat and hydration for babies.
- Do not feed any formula without permission and guidance from a rehabber.
- Use gloves; latex gloves with heavy-duty "cat gloves" are recommended.
- Keep separate from other pets and children.
- Sometimes, skunks get their heads stuck in containers that previously held food items (e.g., yogurt cups, cereal boxes, etc.). If they can't see you, they likely won't spray. Quickly and carefully remove the container.
- Assisting adults can be difficult and hazardous. Utilize humane traps for adults for treatment; avoid handling without experience.

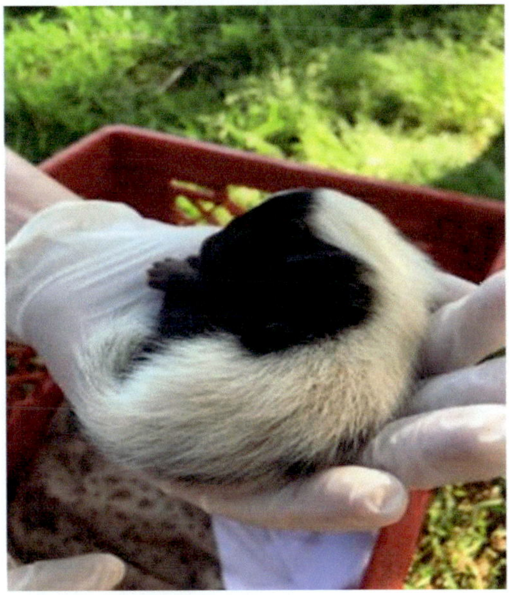

An adorable example of a baby skunk. Still evident what species they are even when they're small due to their unique markings.

Snakes

Normal snake behavior:

- Hunting and feeding on insects, rodents, and sometimes other snakes are natural behaviors that position snakes as essential food chain members. It's important to acknowledge and allow them to follow their natural dietary instincts, even if it might be unsettling for us.
- Different species exhibit varied activity patterns, being either nocturnal or diurnal. Each snake species has its preferred time of day for hunting and other activities.
- Regarding defensive behavior, snakes can appear fearless, and even non-venomous ones can deliver painful bites that might lead to infections. It's advisable to avoid handling snakes unless necessary and to use proper protective gear if you do.
- Snakes often seek shelter in dens in various spots, commonly near water bodies, to regulate body temperature and safety.
- The lifespan of snakes varies by species but can frequently range from 20 to 30 years or more.
- Typically, snake offspring are born in late summer or early fall, aligning with seasonal changes and favorable conditions for their survival.
- Baby snakes can survive independently from the day of birth. No need to intervene if you find a baby.

Abnormal Snake Behavior:

- Presence of flies, maggots, or ants.
- Twitching or thrashing.
- Stuck in netting or on sticky materials.
- Been in another animal's mouth.

Preparing for Rehabilitation:

- Create a calm, quiet, dark, and warm environment for the snake.
- If the snake is awake and aware, soaking it in a small amount of water might be beneficial.
- Refrain from feeding the snake without proper permission and guidance from a wildlife rehabilitator.
- Use protective gloves; latex gloves and heavy-duty "cat gloves" are recommended.
- Keep the snake separate from other pets and children to prevent any accidents.
- Snakes are adept at escaping. Transporting them securely in a pillowcase at the top can be a safe and effective method.
- Assisting a snake in these situations can be challenging and potentially hazardous. Exercise extreme caution and consider seeking professional help if needed. Handling without experience can exacerbate the situation.

Turtles

Normal Turtle Behavior:

- Hunting and feeding on insects, fruits and vegetables, occasionally carrion.
- Different species exhibit varied activity patterns, being either nocturnal or diurnal. You will notice an increase in activity during the cooler parts of the day, usually dawn and dusk, and resting and hiding during the hot parts. Although turtles need sunlight to survive, they don't appreciate the high temperatures.
- The lifespan of turtles varies by species but can frequently range from 20 to 30 years and up to 100 years for our slower-moving land species.
- Depending on the species, eggs usually hatch about 2 months after being laid. Typically, this will occur in late summer or early fall.
- Occasionally, you will see a water turtle out of the water on land. This is normal behavior during the breeding season. Assist across roads, if necessary, in the direction the turtle was headed and leave him to it.

Baby turtles are able to survive independently from the day of birth. No need to intervene if you find a baby. The picture below shows baby snappers the day they're born burrowing into the mud to stay hidden.

Abnormal Turtle Behavior:

- Any swelling around the face, including the eyes and ears, or discharge from the eyes, nose, mouth or ears.
- Cracks, breaks, or bleeding from the shell. If unsure, it's advisable to have a rehabber perform a physical examination before returning the turtle.
- Although turtles can live with missing limbs, having a rehabber to check to see if infection is present in an injury is essential.
- Presents of flies or maggots
- Open-mouth breathing

Note the picture of the turtle with the swollen ears. This indicates an infection that will require the care of a wildlife rehabber.

Handling turtles requires careful consideration. Remember, any creature with a mouth can bite, even the slowest of turtles. Aquatic turtles, in particular, are swift and agile, capable of twisting their necks extensively. Exercise extreme caution when handling them, as their long claws can cause scratches if mishandled.

It's crucial to handle turtles in a way that promotes their comfort. Avoid flipping them upside down or handling them roughly without proper support. Placing a hand underneath their shell can offer stability and reassurance. Keeping them close to your body or within a secure enclosure can also help them feel more at ease. Just a heads up, some turtles' defense mechanism might involve peeing on you, so be prepared.

Preparing for Rehabilitation:

- Create a calm, quiet, dark, and warm environment ideal for turtles.
- If the turtle is awake and alert, soaking it in a shallow water bath might be beneficial to address any maggot infestation. Ensure the turtle can lift its head above water to breathe safely during soaking.
- Refrain from feeding the turtle without proper guidance from a wildlife rehabilitator, as they may not eat during their initial days in captivity.
- Use protective gloves, such as latex gloves, to shield against transmittable diseases.

- Keep the turtle separated from other pets and children to avoid accidents, particularly with other pet turtles.
- Be cautious about the turtle's ability to escape; they are determined and patient creatures, capable of surprising escapes.
- Regarding defensive behavior, turtles, wildly snapping turtles, can appear fearless. It's advisable to avoid handling snappers unless necessary and to use proper protective gear if you do. Use a large snow show or mat to move a snapper out of the road if necessary.

Baby Birds

Handling baby birds can be challenging due to their diverse species and varied needs.

Identifying the baby bird:

Distinguishing between a nestling and a fledgling is crucial. A nestling typically appears pink with fluffy feathers or is naked, bearing no resemblance to adult birds. In contrast, a fledgling resembles a smaller version of an adult bird, sporting most of its feathers. Check back in this book's “renesting” portion to see the difference between the two.

A rehabber should examine a picture of the bird at a minimum before confirming its species. If uncertain, seek advice from someone confident in identification.

If a picture is unavailable, start by weighing the bird; there's a substantial difference between a 50-gram Robin nestling and an 11-gram chickadee nestling. Additionally, maintaining a baseline weight is crucial for your rehabber. It is a significant indicator that the baby consistently gains weight each day, reflecting proper nutrition and care.

Following the weight measurement, observe and note the characteristics of the nest it came from. Take note of the shape and size of the beak, the color of both the beak and the inside of the mouth, and the ratio of fluffy feathers to larger, more mature feathers. These details can assist in identifying or narrowing down the potential species of the bird.

Tips for temporary caging:

Opt for soft-sided cages, ideal for songbirds, to prevent feather damage. Use newspaper lining at the bottom to absorb waste and ensure a dry environment. Include perching materials and consider covering the cage most of the time to reduce stress on the bird. However, it's crucial to note that woodpeckers have barbed tongues, making soft-sided cages risky as their tongues can get stuck and be fatal. For woodpeckers, hard-sided cages without soft bedding are ideal.

Nestlings will require a nest for care. You can repurpose an empty yogurt container or a similar item, adapting it to suit the size of your bird. Line the cup with soft material like cotton or bunting and cover it with multiple layers of toilet paper or tissues. As the bird defecates, remove the soiled layer of tissue and discard it to maintain cleanliness. If the bird repeatedly jumps out of your prepared nest, it may indicate dissatisfaction. Try adjusting the nest to make it deeper or more snug. Your bird should look calm and content while sitting in his nest.

Persistent jumping might also signify it's time for the bird to fledge. Consult a rehabber for guidance in identifying these stages.

Feeding a baby bird:

Diet varies among species, but most baby birds can be nourished with insects.

In urgent situations, most baby birds can survive on mealworms. However, mourning doves, primary seed eaters with tube-like beaks, require immediate attention and tube/crop feeding by a rehabber.

*see the mourning dove picture below and note the shape of the beak.

A healthy baby bird will open its mouth when stimulated for food. Initially, it might not respond to noise as it's unaware of your feeding presence. Gently tapping the side of its mouth can encourage feeding behavior. Over time, as the bird becomes accustomed, it will readily open its mouth. They become increasingly assertive and may have you well-trained to provide food. Use a gloved hand and tweezers to ensure that you keep things sanitary.

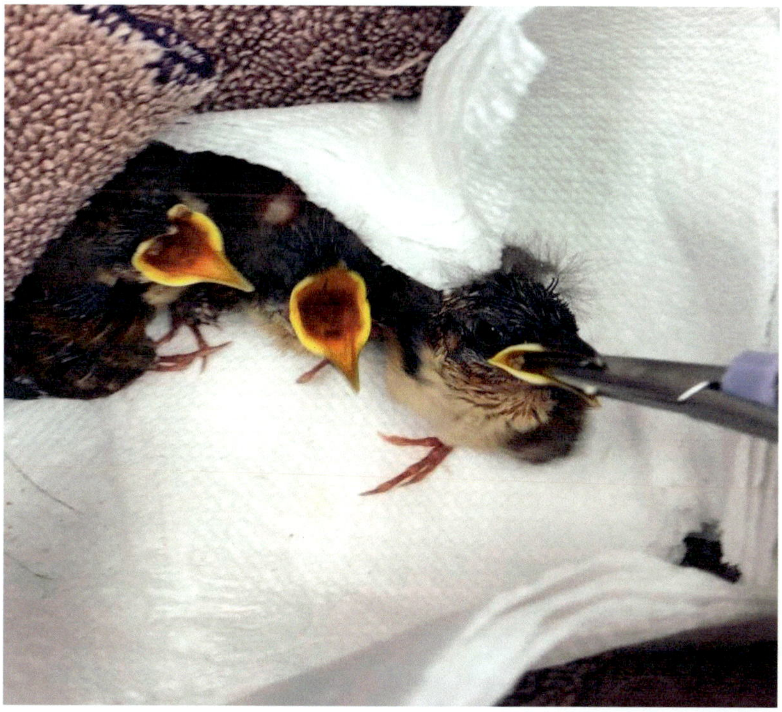

Lastly, before proceeding with the next feeding, check the crop on the side of the bird's neck. The crop should be empty before providing the next meal. If the crop remains full, allow more time before the next feeding. If it doesn't empty after an hour or two, contacting the rehabber immediately is crucial.

Prevent Imprinting

Remember, these birds can imprint easily. Keeping them with others of their species is crucial. Attempting solo rehabilitation might deprive them of vital survival information and social acceptance. While waiting for placement, introducing a hand mirror and feather duster can provide companionship. Please refrain from handling, holding, or cuddling baby birds. Such actions cause undue stress and deviate from their natural development. Training them to perch on your finger or shoulder can set them up for challenges in their natural environment. Always prioritize what's best for them, not what's most comfortable for you.

Chapter 11
Cleaning and Disinfecting After Wildlife

Let's tidy up after aiding wildlife to ensure human health and cleanliness.

Now that you've helped wildlife, you want to keep yourself and YOUR environment safe.

Below is a compilation of recommended disinfectants utilized in our practices for safe wildlife handling. However, these represent just a portion of the available options. It is advisable to conduct thorough research to establish personalized protocols for home use. If there is discomfort in employing solutions such as bleach or potent chemicals, opt for items easily disposed of after use.

Lysol Laundry Sanitizer

We add this to all laundry to disinfect your bedding. You can use bleach, but bleach will deteriorate the fabrics over time. Some centers consider hot water enough to disinfect all but raccoon laundry.

Lysol Laundry Sanitizer is specially designed to sanitize your laundry and to kill 99.9% of bacteria*. It can be used on most washable fabrics, including Baby Clothes, Gym Clothes, Undergarments, Towels, Bedding, and Delicates. *noted on their website

Chlorohexidine Solution

This is used to disinfect after each animal is finished using something. Even if it's been washed, and looks clean, spray it with Chlorahex. Let the object sit for 15 minutes before wiping it off. Most of these disinfectants are light sensitive, so after dilution, store them in an amber-colored spray bottle and give them a proper label.

Chlorhexidine belongs to a group of medicines called antiseptic antibacterial agents. This means it can also be used to clean the skin after an injury, surgery, or injection. Chlorhexidine is also used to clean the hands before a procedure. It works by killing or preventing the growth of bacteria on the skin. Be careful getting Chlorohex around your eyes or using it around young kids. As with any chemical, you should treat it with care.

Kennelsol

This is great for general cleaning and disinfecting. It should be used to soak dishes for 15 minutes after cleaning. Also good for mopping floors.

"KennelSol is a one-step germicidal cleaner, deodorizer and disinfectant specially formulated to safely clean home pet areas and animal care facilities."

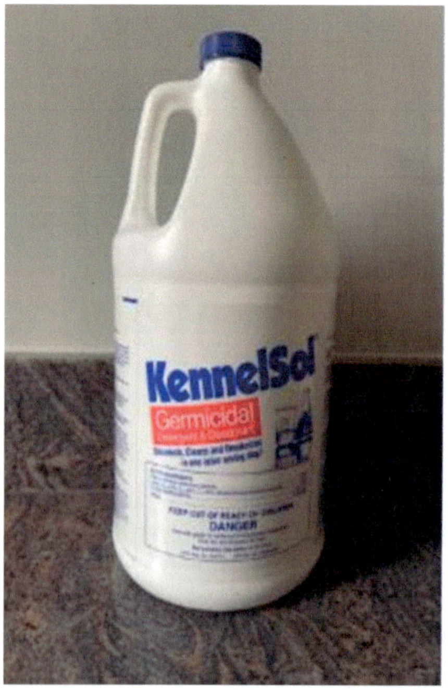

Dawn Dish soap

Use this for general cleaning of dishes and cages, even to clean off toys and climbing equipment in an animal's enclosure. We almost always use a diluted dawn dish soap bath to remove fleas, oils, or any sticky substances. It will take more than one bath to safely remove these residues, but it is worth it over using harsh detergents and chemical insecticides. Warning: Re-read what was just written above. DILUTED dawn dish soap. Never pour dawn directly onto the animal itself. Mix a small amount of the detergent with a lot of water. There should be bubbles, and it shouldn't be dark blue any longer. Look for a more light sky blue concentration. You can also look up specific dilutions online.

"Dishwashing Liquid is both an antibacterial hand soap and dishwashing liquid. When used as an antibacterial hand soap, this unique formula removes 99% of bacteria from hands."

Betadine solution is another great item to keep on hand. Only use it when given guidance by a rehabber or veterinarian.

Don't forget to designate sponges and wash clothes specific to wildlife. You don't want to use the same sponge on your dishes.

Best to label with a Sharpie

How to Do Laundry

Step 1: Ensure all loose debris is shaken off the laundry and put into the trash can. Many centers will give a dry brush to brush off the debris. Over time, if you don't remove the debris, it will break the machine.

Example: dumping all debris into a trash can. Nothing should be left.

Step 2: Always use hot water, laundry detergent, and Lysol sanitizer. Put the sanitizer in the fabric softener compartment. Detergent goes into the detergent compartment. Once you have the detergent in, ensure the temperature is "hot," and the towel setting usually works for my machine. Dry on regular heat.

Step 3: Remember that when in doubt, throw it out! If you have a messy situation or an animal shreds a particular item pretty severely, don't spend too much time trying to fix it. Sometimes, it's ok to throw it out.

Chapter 12
Make a Difference in The World by Making a Difference in Your Community

In our world, remarkable individuals are engaged in beautiful endeavors to foster harmony between humanity and wildlife. It remains crucial to acknowledge that each of us holds the power to contribute to this meaningful cause. It is our responsibility to honor the very land we inhabit. Often, these contributions to our daily routines may seem modest. Yet, these seemingly trivial actions accumulate, significantly aiding the realization of larger-scale initiatives. Sometimes, the most substantial contribution lies in simply taking that initial step and seeing a task through to completion. Regardless of its scale, for, as the adage goes, 'A journey of a thousand miles begins with a single step.

Tanya's Turtle Project

In my experiences with wildlife, I've encountered numerous individuals who've deeply inspired me and the broader movement for wildlife. It's remarkable how a single spark can ignite an entire community's engagement. Sometimes, an idea strikes that's so uniquely profound, carrying the potential to make an immense difference, and it has the power to inspire people from all walks of life to join hands in accomplishing it.

One such luminary in the world of wildlife rehabilitation is Tanya Finch. Her journey began with an encounter in 2017 while walking her dogs near a construction site. Her dogs found space to play and run laps inside the silt fencing. While walking in that expanse, Tanya spotted a Woodland Box Turtle trapped in the mud, unable to navigate the barrier. Woodland Box Turtles, ill-equipped to overcome such obstacles, often end up in futile circular paths along fencing, ultimately falling victim to construction or remaining trapped.

That day, Tanya rescued 11 Woodland Box Turtles, lifting them over the silt fencing. This event ignited her passion project, Tanya's Turtle Project.

Delving deeper into research, Tanya uncovered Woodland Box Turtles' dire predicament. These creatures can live up to 80 years, only reaching sexual maturity after 15 to 20 years. This is much the same as a human lifespan. As a female Woodland Box Turtle ages, her ability to produce more and healthier eggs increases, potentially ensuring the species' survival. Saving just one adult female could mean the difference between sustaining Woodland Box Turtle populations or their disappearance from an area.

The decimation of their habitats due to human activities exacerbates this crisis. Deforestation for urban expansion drastically limits their space. Mere pockets of greenery amidst concrete landscapes cannot sustain diverse wildlife; it takes ample resources to support each species. Centuries of survival instincts honed in the wild are jeopardized when confined to fragmented, inadequate habitats.

This struggle isn't exclusive to woodland box turtles; Our aquatic turtle species confront comparable challenges. Our negligence has led to severe pollution of waterways. Additionally, we've constructed new water bodies and retention ponds where turtles often establish their homes and nesting sites. Yet, every few years, the need arises to drain these ponds, reconstruct them, and build anew to accommodate the ever expanding human communities around these areas. How can these turtles secure future generations when their habitats are continually bulldozed and disrupted?

Additionally, Each year, hundreds of Woodland Box Turtles are admitted to rehab facilities in Virginia, victims of other various human-inflicted, although sometimes indirectly caused, injuries. From rodenticides to habitat loss, dog attacks and vehicle strikes, their population is under siege. Tanya and her project offer a glimmer of hope for these creatures.

It was a pleasant surprise when she connected with construction contractors, who were supportive and eager to help her move the project forward and find outlets for its progress. This experience highlighted that sometimes, we shouldn't hastily assume that people don't care; often, they genuinely want to assist but might not know where to get started.

Tanya's Turtle Project has grown into many wildlife advocacy groups collaborating closely with construction company leaders and property owners overseeing construction sites. Their joint efforts involve patrolling the grounds to assist animals attempting to navigate away from construction zones. Tanya's initiative has shed light on these animals' challenges, inspiring the public and fostering a sense of empowerment that each individual can effect change.

We're not requesting the relocation of these animals or the construction of makeshift shelters. Instead, as you pass the silt fencing encircling a construction site, we ask for a few minutes of your time. If you happen upon any creatures trapped in the mud, striving to escape, a simple act of assistance—picking them up and safely placing them on the other side of the fencing—could make an enormous difference in their survival.

It's easy to dismiss these actions as inconsequential, but they could alter the fate of generations of Woodland Box Turtles. A simple act like assisting a female across a barrier or aiding them across a perilous road might ensure the survival of an entire lineage.

Tanya Finch and her project highlight the challenges these creatures face. They show everyone that even little things can make a big difference in preserving these incredible animals, empowering all of us to take action!

Bill Crisp and Snakes Education For the Public

When discussing wildlife advocates, another notable figure comes to mind: Bill Crisp with K2C Reptiles. He has dedicated his efforts to collaborating with the public, assisting them in overcoming their fear of snakes and shedding light on how to coexist harmoniously with this unique reptilian group.

Like any other creature, snakes play a vital role in our ecosystem and are essential members of our wildlife community. Unfortunately, they are often the first to be targeted based on their presence. Many people harbor an aversion to snakes, finding their sight alone fear-inducing and perceiving them as a threat. Common sentiments express a desire to kill snakes upon immediate encounter, believing that "the only good snake is a dead snake." However, these notions are not accurate.

Having snakes around actually offers excellent benefits. Firstly, they serve as highly effective pest control. If snakes are present, it often indicates a rodent issue in the area. Where there are mice in the house, a snake is likely to follow. It's important to recognize that the snake is performing a valuable service by helping to keep the rodent population in check. This balance is essential for maintaining biodiversity in a healthy ecosystem.

Beyond pest control, snakes have become valuable in medical research. Their venom is now being utilized in the development of medications, including anticoagulants, painkillers, and even for breast cancer research. The venom of snakes holds the potential to create innovative products that could contribute to saving lives. Therefore, acknowledging and appreciating the role of snakes in our environment is crucial for both ecological balance and scientific advancements.

The most effective approach to facilitate the safe adaptation of snakes to our environment is through public education, particularly targeting children, to raise awareness about the benefits of these creatures.

This is precisely the mission that Bill Crisp has undertaken. Over time, Mr. Crisp observed a widespread hesitancy towards snakes, often stemming from ignorance and an inability to accurately identify them. To address this issue, he began fielding phone calls and text messages to assist the public in snake identification.

This initiative sparked a movement in Northern Virginia, inspiring and empowering individuals to coexist harmoniously with snakes. Bill Crisp has taken it a step further by offering classes to the public, providing a comprehensive understanding of Virginia snakes. In these sessions, he introduces his own pet snakes, offering attendees the opportunity to witness them in a controlled environment and encouraging them to touch and hold these remarkable creatures, which have grown comfortable with human interaction.

Such advocacy goes beyond dispelling fear; it actively counters the fear factor contributing to the decline of snake species. By fostering understanding and firsthand experiences with these animals, Bill Crisp's educational efforts are crucial in reshaping perceptions and promoting coexistence between humans and snakes.

Recognizing the types of snakes in your surroundings is crucial, but distinguishing between venomous and non-venomous species is equally vital. This skill has proven to be a valuable tool for numerous homeowners. Simply being aware of the presence of venomous snakes in your vicinity is a starting point for implementing practices that help prevent bites and promote peaceful coexistence. This knowledge empowers individuals to take proactive measures, fostering a safer environment and enabling a more informed and harmonious interaction with the diverse snake population in their area.

Given the prevalence of copperheads in our area, Bill addresses their behavior patterns in his discussions. He emphasizes their occasional nocturnal nature and often engages in interviews with individuals who have experienced copperhead bites, aiming to gain a deeper understanding of the circumstances surrounding the incidents. One recurrent observation is the significance of wearing appropriate footwear, a simple yet effective precaution.

Through these interviews, Bill identified a common scenario: many people take out their trash on warm evenings without wearing shoes. The sudden movement and disturbance of the trash can provoke a defensive strike from copperheads toward exposed feet. This seemingly minor detail can lead to significant consequences. Therefore, the advice is clear: wearing shoes, even during routine activities like taking out the trash on warm evenings, can substantially reduce the risk of snake bites. These small, practical measures can make a considerable difference in promoting safety and minimizing potential hassles.

The Importance of Wildlife Rehabilitation in Your Community

Over the years, wildlife rehabilitators in Virginia have encountered many animals. A.E.R.O. Animal Education and Rescue Organization. Animal Rescue, a non-profit group of at-home rehabbers, has consistently handled over 1000 animals annually for the past three years. Serving as the boots on the ground, we offer a firsthand perspective on the overall wildlife scenario. Wildlife rehabilitators must take responsibility for the information they gather by meticulously documenting and analyzing it for future research and reference.

As the initial responders, wildlife rehabilitators play a crucial role in observing and understanding population trends and species distribution throughout the state. We witness the fluctuations in these patterns, and it raises concerns when we observe declines. Furthermore, wildlife rehabbers are often the first to detect patterns or noteworthy wildlife illnesses or injuries, enabling the identification of emerging issues with implications for public health.

In 2021, the emergence of Brood X cicadas after a 17-year hiatus coincided with an alarming phenomenon: birds entering rehabilitation with crusty swollen eyes and neurological symptoms. Faced with an unknown issue for which there was no established treatment, many of these birds succumbed to the condition or had to be euthanized. In areas without accessible wildlife rehabilitation, such occurrences might have gone unnoticed by many.

Fortunately, Northern Virginia boasts a stable group of wildlife rehabilitators. When rehabbers across different counties simultaneously encountered calls and intake with similar symptoms, it provided an opportunity for collaboration with scientists to help unravel the mystery. Proper care, training, and precautions were crucial in evaluating these affected birds, coupled with the necessary experience in handling songbirds to discern what is normal and abnormal within these species. Certain viruses and infectious diseases are more common in songbirds, but it takes real-world experience to distinguish between well-studied and understood conditions and novel, emerging issues that may carry dangerous implications. This collaborative effort among wildlife rehabilitators and scientists highlights the importance of a proactive and knowledgeable response to protect avian populations during unforeseen events.

Wildlife admitted to rehabilitation facilities typically adheres to seasonal patterns corresponding to offspring birth. Through years of experience, Rehabbers become adept at recognizing these patterns and can anticipate the specific month when particular species of animals are expected. With ongoing environmental changes, rehabbers are noticing shifts in admission patterns indicative of broader environmental changes. Systematically documenting these observations and sharing the information with other science-based organizations can provide valuable insights into the state of our local environments. This collaborative effort contributes to a deeper understanding of regional ecosystems and has broader implications on a global scale. By connecting these localized observations, scientists and environmentalists gain a more comprehensive perspective on the impact of environmental changes, helping inform conservation strategies and promote sustainable practices both locally and internationally.

Moreover, wildlife rehabilitators offer firsthand accounts of the challenging interactions between wildlife and humans. They witness the unfortunate consequences of injuries resulting from collisions with vehicles, habitat destruction, encounters with domestic animals, and the implementation of unsafe practices to address wildlife issues in domestic areas. It is, therefore, crucial for rehabbers to actively engage with the public and provide guidance on alternative approaches to navigate wildlife conflicts in the future. At the same time, these situations may be disheartening and present an unsightly side of human-wildlife interactions, using them as learning experiences provides valuable education opportunities. Wildlife rehabilitators contribute to public awareness and understanding by sharing insights gained from these encounters. Turning these challenges into educational moments ultimately benefits wildlife by fostering a greater appreciation for coexistence and promoting responsible practices in human-wildlife interactions.

Being a wildlife rehabilitator entails considerable responsibility, extending beyond the care and compassion required for animal rehabilitation. It required time and energy to collaborate with peers and share acquired knowledge. Presenting this information to the wider community contributes to improving best practices globally. Encouraging wildlife rehabilitators to step forward and share their insights can be challenging, as practices may face criticism from the public and fellow rehabbers. What's essential for people to understand is that there is no one-size-fits-all approach to wildlife rehabilitation. Each practice can vary from the techniques used to feed a baby animal to the choice of disinfectants. Patience is key, and fostering an environment of mutual encouragement is crucial. Rehabbers should support one another in adopting care techniques that work, recognizing that the ultimate measure of success is the positive outcome achieved in their specific facility.

Encouraging and supporting wildlife rehabilitators through various means, including donating time, money, and resources and expressing gratitude, is important. In a world quick to highlight shortcomings, acknowledging and celebrating the positive impact of their work is equally important. By reinforcing a supportive community, we can inspire continued dedication to wildlife rehabilitation and highlight the positive contributions made by these caring individuals.

Now Go Out and Enjoy The Wild World

Finally, let's delve into the beauty of appreciating wildlife. Amid the ongoing discussions about the various challenges confronting our world, especially those related to the environment, it's important to pause and reflect on the significance of preserving it. Nature exists for our coexistence, and as we deepen our knowledge about it, we foster a profound understanding of how to truly appreciate it and what it truly is. It's magic. When individuals isolate themselves from the natural world, they miss out on the many wonders nature has bestowed upon us. Take a stroll outdoors, embark on nature walks and hikes—explore the local parks and nature preserves with trails for observing wildlife. Familiarize yourself with the diverse fauna and flora in your area. Engage in birdwatching and be astonished by the wealth of knowledge you can acquire. Each species is unique, akin to meeting a new person—unfolding new personalities and tales to share. Capture these moments through photography, share your experiences, and converse with others about the wonders you encounter. We must shift our focus toward highlighting the positive aspects of the world around us. If more attuned to the benefits, we would treat our environment with greater care and consideration.

Make a positive impact on the world by getting involved. Nothing is going to change if nothing ever changes. Be the driving force in your community that ignites a passion for the wonders of nature. After all, they say no remarkable story originates from sitting on your couch. Step outside and discover where the journey leads you. Perhaps today is the day you save an animal's life, or maybe it's the day it saves you.

Made in the USA
Monee, IL
05 November 2026

67387142R00057